STUDENT UNIT GUIDE

NEW EDITION

Edexcel A2 Geography Unit 3
Contested Planet

Cameron Dunn and Sue Warn

Philip Allan Updates, an imprint of Hodder Education, an Hachette UK company, Market Place, Deddington, Oxfordshire OX15 0SE

Orders
Bookpoint Ltd, 130 Milton Park, Abingdon, Oxfordshire OX14 4SB
tel: 01235 827827
fax: 01235 400401
e-mail: education@bookpoint.co.uk
Lines are open 9.00 a.m.–5.00 p.m., Monday to Saturday, with a 24-hour message answering service.
You can also order through the Philip Allan Updates website: www.philipallan.co.uk

ISBN 978-1-4441-4767-4

First printed 2012
Impression number 5 4 3 2 1
Year 2015 2014 2013 2012

Cover photo: Corel

Printed in Dubai

Hachette UK's policy is to use papers that are natural, renewable and recyclable products and made from wood grown in sustainable forests. The logging and manufacturing processes are expected to conform to the environmental regulations of the country of origin.

P01952

Contents

Content Guidance

Questions & Answers

Getting the most from this book

Examiner tips

Advice from the examiner on key points in the text to help you learn and recall unit content, avoid pitfalls, and polish your exam technique in order to boost your grade.

Knowledge check

Rapid-fire questions throughout the Content Guidance section to check your understanding.

Knowledge check answers

1 Turn to the back of the book for the Knowledge check answers.

Summary

Summaries

- Each core topic is rounded off by a bullet-list summary for quick-check reference of what you need to know.

Questions & Answers

Exam-style questions

Examiner comments on the questions
Tips on what you need to do to gain full marks, indicated by the icon **e**.

Sample student answers
Practise the questions, then look at the student answers that follow each set of questions.

Examiner commentary on sample student answers
Find out how many marks each answer would be awarded in the exam and then read the examiner comments (preceded by the icon **e**) following each student answer. Annotations that link back to points made in the student answers show exactly how and where marks are gained or lost.

About this book

This guide is for students following the Edexcel A2 Geography course. It aims to guide you through **Unit 3: Contested Planet**. The **Content Guidance** section provides a detailed guide to the six topics that make up the 'Contested Planet' unit. The **Questions & Answers** section provides information on assessment and synopticity, as well as examples of six Section A questions, with outline mark schemes, examiner's comments and contrasting student responses with commentaries. The examiner's comments indicate how the question should be approached and how the answers could have been improved.

Overview of Unit 3

Unit 3 examines a range of contemporary geographical issues that all relate to the idea of a contested planet. It is divided into the following six topics:

Topics 1–3: Energy security, Water conflicts and Biodiversity under threat — focus on resources. The growing human population increasingly puts resources under pressure, leading to conflicting views over their use and management.

Topics 4 and 5: Superpower geographies and Bridging the development gap — investigate the two faces of development: the wealthy and powerful versus those on the 'wrong side' of the development gap.

Topic 6: The Technological fix? — uses the theme of technology to explore different ways in which global problems and issues might be managed to create a more sustainable planet.

Each of the six content sections of this book follows the specification closely. Each section is divided into three subsections as in the specification. In the specification, the left-hand column is what you need to learn — this is the one the examiners will use when setting questions in the exam. The right-hand column suggests how you might study the concepts and gives you a context for your studies.

The Content Guidance section summarises the key information you need to know for each of these topics. The information for each topic is divided into subsections that match those of the specification.

At the end of each topic, there is a synoptic links section. This is designed to help you see how each topic relates to other areas of geography, both at AS and A2. The synoptic links sections will be useful when you come to prepare for the Unit 3 Section B issues analysis.

Content guidance

Energy security

Energy supply, demand and security

Types of energy source

Energy is fundamental to human existence. We need energy for transport, heating, cooking food and surfing the web. Eighty-five per cent of global energy consumption in 2007 was from fossil fuels (coal, oil and gas). Our dependence on fossil fuels is remarkable because extensive use of fossil fuels only began a few hundred years ago. Before the Industrial Revolution most energy sources were renewable (Table 1), such as water wheels, windmills and wood (biomass).

Table 1 Energy sources and percentage of global energy supply

Renewable	Non-renewable	Recyclable
• Wind turbines (0.3%) • Solar photovoltaic cells and passive solar (0.5%) • Wave power • Tidal power • Hydroelectric power (HEP) (3%) • Geothermal (0.2%)	• Coal (25%) • Oil (37%) • Gas (23%) • Unconventional oil and coal, e.g. tar sands, heavy oil, oil shale, lignite and peat	• Biomass (4%) • Biofuels (0.2%) • Nuclear power (with reprocessing of fuel) (6%)
These resources result from a flow of energy from the sun or Earth's interior	There is a finite stock of these resources, which will run out	These resources have a renewable stock, which can be replenished with careful management

Knowledge check 1

Why is nuclear power classed as a recyclable energy source?

The environmental consequences of using the three energy sources in Table 1 are quite different:

- Renewable sources produce no carbon dioxide, and do not directly contribute to atmospheric pollution.
- Non-renewable sources emit carbon dioxide during combustion, and are the cause of global warming.
- Recyclable biomass and biofuels emit carbon dioxide, but reabsorb it when they are regrown — making them potentially close to being 'carbon neutral'.

Examiner tip

You need to be able to compare the environmental consequences, both positive and negative, of using all of the energy sources in Table 1. You could complete an A3 size table to revise this.

Nuclear power stations do not emit carbon dioxide but there are significant environmental concerns about the radioactive uranium fuel, and the long-term problem of disposing of nuclear waste.

Distribution

Physical geography largely determines direct access to energy resources. Whether coal, oil or gas are beneath a country is an accident of millions of years of geological processes. By chance, the UK ended up with rich reserves of coal, oil and gas. With large tidal ranges of up to 15 metres, some of Europe's strongest winds and many glacial valleys, the UK has significant renewable resource potential. However, high latitude areas like the UK have low solar power potential. Tectonically active areas such as Iceland have much higher geothermal potential. Energy resources are concentrated geographically:

- In 2005, four countries made up over 70% of global uranium production for nuclear power — led by Canada (28%) and Australia (23%).
- By 2025, 60% of world oil supply will come from the Middle East.
- Currently, 27% of all proven natural gas reserves are in Russia.

Examiner tip

Make sure you understand the difference between total energy use from all sources and energy sources used to generate electricity.

Geography means some countries have vast energy surpluses (Russia, Saudi Arabia) but others suffer from energy poverty. Increasingly there is a mismatch between demand for fossil fuels and supply (Figure 1). Some countries such as Mali appear to have no fossil fuel reserves. Mali has huge solar potential, but solar cell technology is very costly despite the energy source being 'free'.

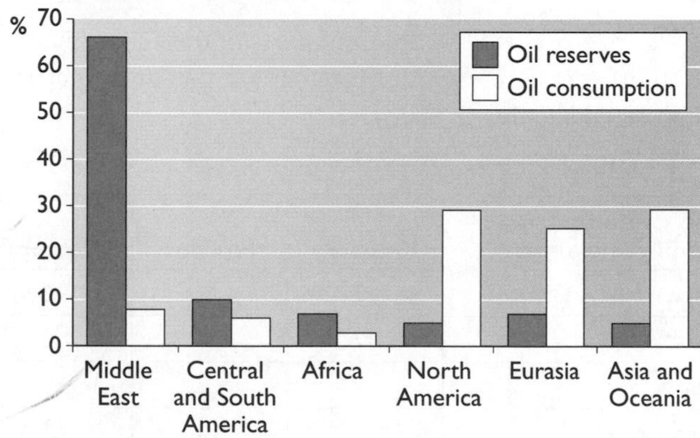

Figure 1 Oil reserves and oil consumption in 2005 by global region (% of global)

Knowledge check 2

Which region is the most significant in terms of the global supply of crude oil?

Energy use

The energy 'mix' a country chooses results from a number of factors.

- Physical — the availability of North Sea natural gas contributed to a 'dash for gas' in the early 1990s.
- Public perception — in the 1950s and 1960s nuclear power was perceived as a positive technology, but after the Chernobyl disaster in 1986 the public turned against nuclear power.
- Politics — nuclear power is back on the agenda due to fears over the politics of gas supply from Russia, although the 2011 Fukushima disaster could be a set-back.
- Technology — solar panels efficiency increased from 5% energy conversion to 40% energy conversion between 1970 and 2008, increasing its viability.
- Economics — wind power is becoming competitive with fossil fuels. In the USA the installed cost is around US$55 per MWh — almost the same as coal and gas.

- Environment — concerns about global warming have led a move towards renewable sources. UK wind generating capacity increased from under 500 MW in 2001 to 5200 MW by 2011.

As Figure 2 shows, UK total electricity generation has grown but there has been a significant shift to gas since 1987, and a decline in nuclear power since 1997 (due to closure of old power stations, and no replacements).

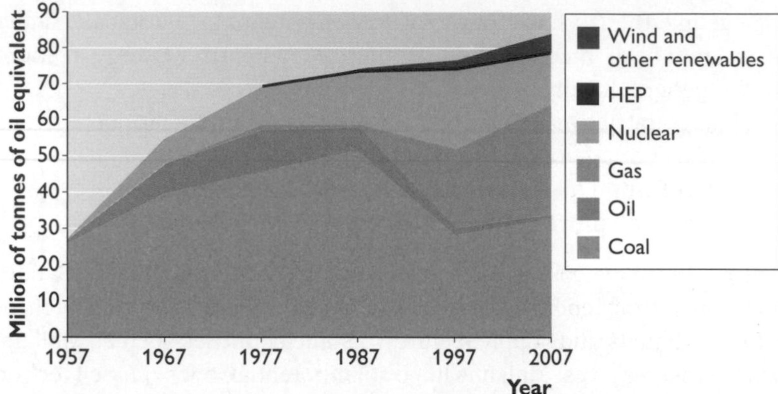

Figure 2 The changing UK energy mix for electricity generation

Different energy sources are used by different parts of the economy (Table 2), with oil dominating transport (petrol, diesel and fuel oil) but gas and electricity dominating domestic and industrial supply.

Examiner tip
Learn some key energy statistics for the UK, including trends in the use of different energy sources.

Table 2 UK energy use by source and sector in 2006

2006 Million tonnes of oil equivalent	Industry	Homes	Transport
Coal	2	1	0
Gas	12	31	0
Oil	7	3	59
Electricity	10	10	1

Energy trends

Energy demand globally is projected to grow by as much as 50% between 2005 and 2030 (Figure 3).

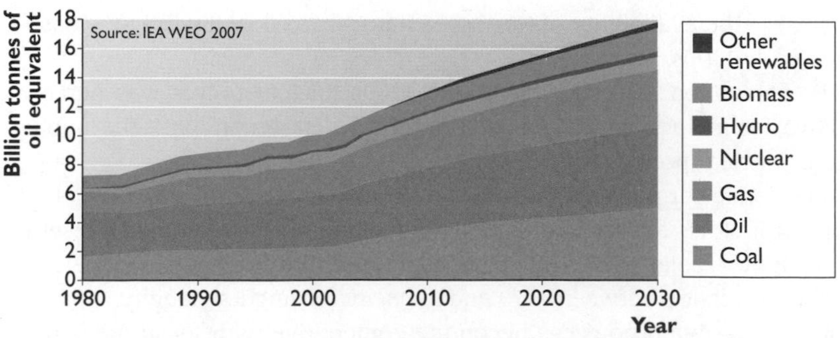

Figure 3 Global energy demand by type 1980–2030

Growth is expected to be 0.7% per year in developed countries but 2.5%+ in emerging economies such as India and China. In China, coal use increased by 17% per year between 2002 and 2005 as the country tried to meet its insatiable demand for power. Most projections suggest a continued reliance on fossil fuels, rather than a switch to nuclear power and renewables.

This is because:

- nuclear plants are costly to build, and take up to 10 years to complete
- renewable sources such as wind are seen as unreliable
- solar power has proved difficult to 'up-scale'
- China and India have both 10%+ of world coal reserves, and coal power stations can be built cheaply and quickly

If projected energy demand materialises, it will have implications for the price of fossil fuels, plus major environmental consequences.

Knowledge check 3

What is the importance of the BRIC economies in terms of global energy demand?

Energy security

A key issue for China and India is security of energy supply. We all expect to be able to get the energy we need, at an acceptable price. The ability to do this is energy security, which depends on the factors shown in Figure 4.

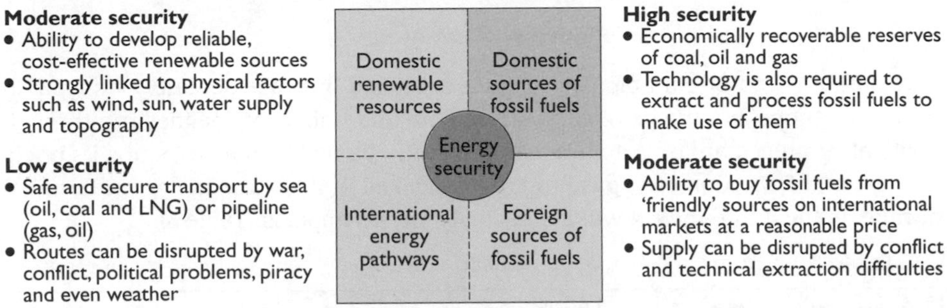

Figure 4 The energy security square

Countries dependent on foreign sources of fossil fuels transported along international energy pathways, are at greater risk of energy insecurity. This explains why some countries, notably France (86% of electricity supply) and Japan (30% of electricity supply), have invested heavily in nuclear power to reduce dependency on imported fossil fuels.

Examiner tip

Learn a definition of energy security for the exam, which should refer to domestic energy sources, potential, foreign sources and energy pathway security.

The impacts of energy security

Energy pathways

As remaining fossil fuel reserves become concentrated in only a few countries, the pathways used to transport oil and gas around the globe will become more significant. Figure 5 shows some key energy supply hotspots.

Western Europe is increasingly dependent on Russian gas. This has given Russia new political power, and foreign currency. Gas export pipelines pass through former

Soviet republics such as Ukraine and Belarus. In 2006 and 2009, gas supplies to Ukraine were cut off over payment and price disputes, causing supplies down-line in France and Germany to fall by 20–30%.

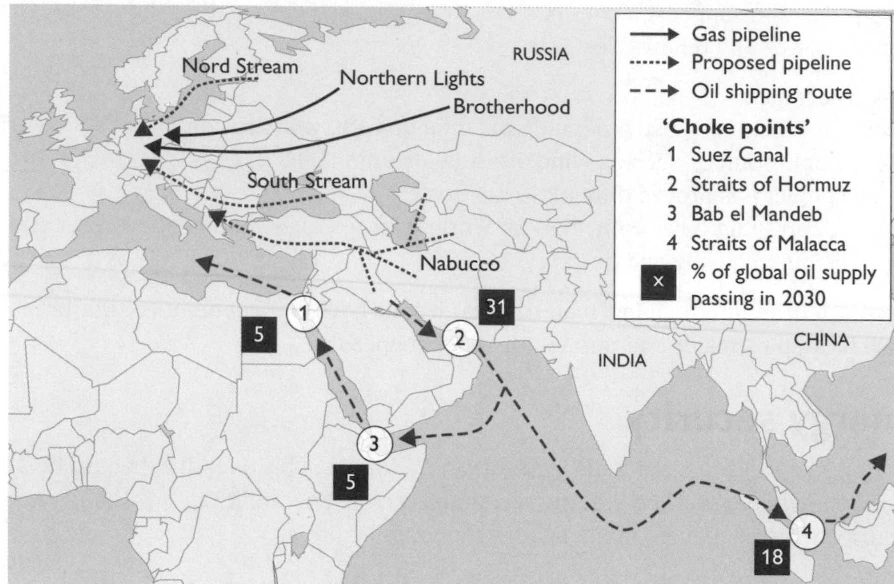

Figure 5 Energy hotspots

Examiner tip

Remember that different energy sources are transported along pathways in different ways, i.e. electricity via cables, gas and oil by pipeline or ship, and coal by ship, road or rail.

The planned Nord Stream pipeline (Figure 5) should increase security of supply in northern Europe, but the planned South Stream and Nabucco pipelines run through politically troubled areas. Europe's fear is that Russia will be able to 'name its price' for gas. In the UK, domestic gas production peaked in 2000 and is now in terminal decline. In the future the UK will increasingly rely on imports (Table 3).

Table 3 A possible UK natural gas scenario for 2015 (UK Energy Research Centre)

UK projected gas demand in 2015 = 123×10^9 m^3 By 2015 77% of UK gas will be imported	UK domestic production	32×10^9 m^3
	Norwegian imports	33×10^9 m^3
	Russian imports	44×10^9 m^3
	Middle Eastern liquefied natural gas (LNG)	34×10^9 m^3

By 2030, over 30% of world oil is likely to pass through the narrow Straits of Hormuz in the Persian Gulf. Historically this is an area of conflict (Iran–Iraq War 1980–88, Gulf War 1990–91, 2003 Invasion of Iraq). Oil and gas pipelines and supertankers are vulnerable to attack during war and from terrorism. Supertankers are vulnerable to piracy. Somali pirates seized the supertanker MV *Sirius Star* in 2008. The Straits of Malacca are a piracy hotspot. There is growing concern over the volume of oil that passes through so-called narrow 'choke points' and the ease with which these could be disrupted or blocked.

Knowledge check 4

What is the significance of choke points to the global oil supply chain?

The costs of disruption

The risks of disruption to energy supplies can be seen by examining past oil prices (Figure 6). Since 1970, the oil price has spiked four times and each time a period of economic recession has followed.

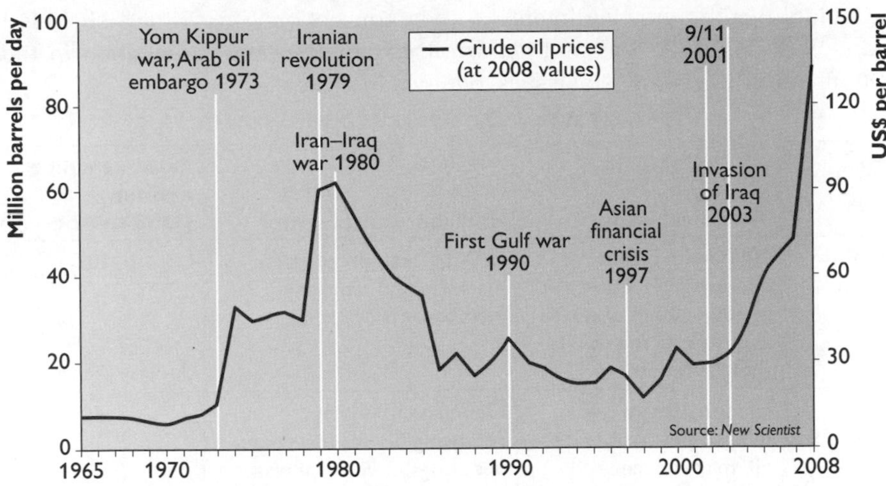

Figure 6 The world crude oil price 1972–2008

High oil prices increase costs for industry, which leads to inflation and rising prices. People spend proportionately more on expensive energy and less on other goods. This slows, or reverses, economic growth. As energy is so fundamental to the developed world, high energy prices increase political and economic risks.

- High petrol prices in 2007–08 contributed to a collapse in sales of gas-guzzling SUVs in America, and major difficulties for Ford, Chrysler and General Motors.
- In the UK in 2000, petrol price rises led to a series of protests and blockades that forced 3,000 petrol stations to close.
- In 2008, 70,000 angry lorry drivers blockaded the Franco-Spanish border leading to panic buying in the shops, as supplies of food dwindled.

Examiner tip

Keep an eye on the price of oil, which is increasingly volatile. You can find oil price data on the BBC business news website.

These events show the importance of energy supply. People will take to the streets if they feel they cannot get, or afford, energy.

Governments can get energy policy wrong. Since 2007, South Africa has suffered from periodic power blackouts referred to by the national electricity company, Eskom, as 'load shedding'. Eskom does not have enough generating capacity to meet demand — a result of years of under-investment. In 2008:

- gold and platinum mines cut back production, or shut, due to electricity shortages
- foreign direct investment was reduced due to supply fears
- estimates suggest 2% was trimmed from economic growth
- blackouts caused traffic chaos and closed shopping malls

Knowledge check 5

Why are developed world consumers sensitive to the price of oil, gas and electricity?

South Africa is planning new power stations, but these take time to build and Eskom has warned of continuing 'load shedding' into 2012.

Looking for more energy

The search for secure oil and gas supplies has led to exploration in technically difficult areas:

- extreme cold environments, such as inside the Arctic Circle
- deep water offshore locations such as the west of Shetland
- politically unstable locations such as Sudan and Puntland in northern Somalia

Examiner tip

You could research the growth of shale gas production in the USA, as this is a growing, but controversial, energy resource.

In addition, there is growing interest in non-conventional fossil fuels such as tar sands, oil shales, shale gas and heavy oil. Exploration and exploitation in these locations raises environmental issues, outlined in Table 4.

Table 4 The new oil frontiers

Location	Oil type (+ recoverable reserves)	Technical issues	Environmental issues	Price at which economic (US$ per barrel)
West of Shetland (Foinaven) oil field, UK, discovered 1992	Light oil (250–600 million barrels) and gas	Oil pumped from floating production ship in 400–600 m of water and transferred to shuttle tanker	Lack of fixed production platform and pipelines increases risk of spill	10+
Athabasca tar sands, Canada, covering 140,000 km^2	Sand and bitumen (tar) mix (170 billion barrels)	Injecting steam into the ground to liberate oil, may reduce environmental impact	Opencast mined by removing boreal forest and peat bogs. 2–5 m^3 of water used for 1 m^3 of oil; natural gas used to heat tar sands and recover oil	40
Orinoco heavy oil, Venezuela	Heavy oil sands (200+ billion barrels)	Challenging physical geography and difficult to transport oil; much is shipped as an oil-and-water mix called orimulsion	The Orinoco river and delta are fragile, biodiverse tropical environments which could easily be damaged by oil exploitation	40
Green River Basin, USA, oil shales	Sedimentary rock containing kerogen (750 billion barrels?)	Likely to be more viable and acceptable if an in situ extraction method is perfected	Surface mining has the potential to cause major environmental damage; acid runoff a likely side-effect; dirty fuel if burnt directly	80–100

Knowledge check 6

What are the main arguments against developing unconventional fossil fuels such as those shown in Table 4?

In the last decade attention has turned to the Arctic as a possible location of major new oil and gas reserves. Estimates in 2008 suggested that the Arctic Ocean could contain 90 billion barrels of oil, and as much as 30% of the world's undiscovered natural gas.

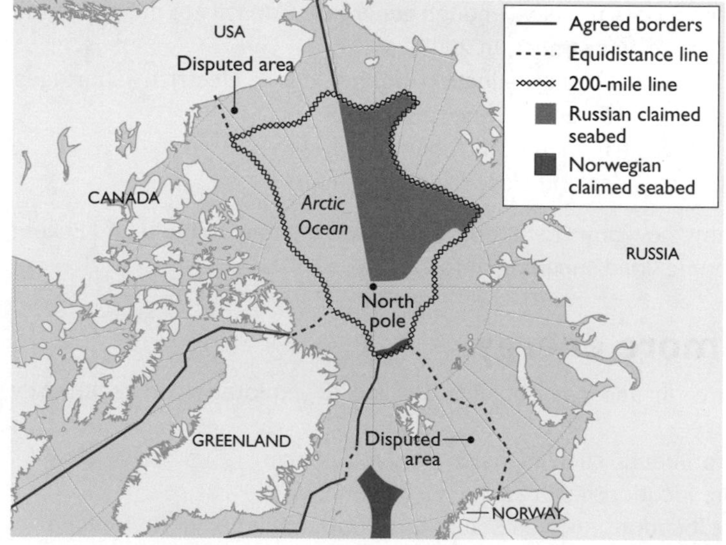

Figure 7 Oil in the Arctic

Source: Durham University, UN, Marum

The whole area is subject to territorial claims (Figure 7), which have yet to be settled by the UN. Environmentalists fear an Arctic 'free for all' as fossil fuel reserves elsewhere run dry, or worse, the possibility of conflict if territorial claims are not settled. The 2010 BP Deep Water Horizon oil spill in the Gulf of Mexico illustrated the dangers of exploring for oil at the technological frontier. Eleven people were killed when the rig exploded and over 4 million barrels of oil were spilled.

Examiner tip

You studied the Arctic in AS Unit 1. Revisit your notes to remind yourself about the fragility and vulnerability of Arctic ecosystems.

Players in the energy game

Energy is big business. Very large TNCs known as 'supermajors' and equally large state-owned oil and gas companies (Table 5) dominate.

Table 5 Supermajor oil and gas TNCs, and state-owned companies

Supermajors (2009/10 revenue in billions US$)		State-owned companies (2009/10 revenue in billions US$)	
ExxonMobil (USA)	383	Saudi Aramco (Saudi Arabia)	233
Royal Dutch Shell (UK/Neth)	355	CNPC (China)	165
BP (UK)	309	PDVSA (Venezuela)	91
Chevron Corporation (USA)	205	Petrobras (Brazil)	138
ConocoPhillips (USA)	199	Gazprom (Russia)	99
Total (France)	231	Petronas (Malaysia)	66

These companies are involved in exploration, extraction, refining and delivery. They are vertically integrated TNCs involved in the entire energy supply chain. Many are diversifying into renewable energy, hedging against the time when oil and gas run out. Supermajors have been heavily criticised for:

- making excessive profits; many are close to being monopoly suppliers in some countries and regions
- not investing long term in exploration and refining capacity, so oil gluts quickly turn to supply shortages
- damaging sensitive environments and ignoring local people, such as in Rivers State in Nigeria where Shell has been accused of oil spills, corruption and driving the Ogoni people from their land

Examiner tip

Visit a supermajor website such as **www.shell.com** or **www.bp.com** and examine the global nature of their business and any evidence of diversification away from fossil fuels.

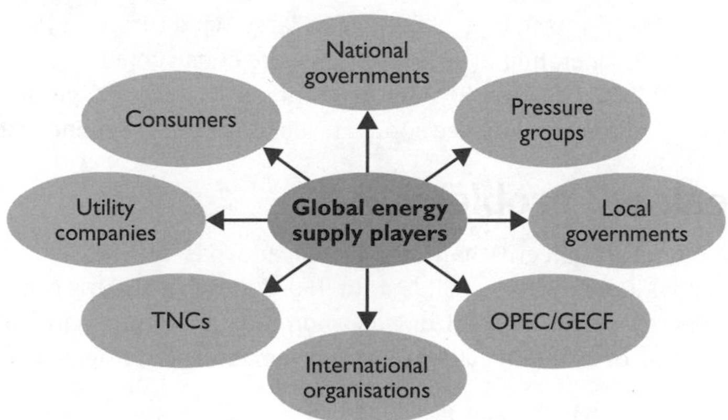

Figure 8 Players in the global supply of energy

OPEC (the Organization of the Petroleum Exporting Countries) is a powerful energy player. Set up in 1961 it is a cartel of 12 major oil exporters. Members include Iran, Kuwait, Saudi Arabia, Venezuela and Nigeria. OPEC influences the price of crude oil using oil production quotas for member states. These can decrease oil supply, increasing prices. In 2008, OPEC accounted for 35% of oil supply, but over 60% of proven oil reserves. Its power is likely to increase in the future. The less influential Gas Exporting Countries Forum (GECF) was established in 2001. There are concerns that Iran and Russia may attempt to make GECF into a price setting cartel for natural gas.

Knowledge check 7

Explain the importance of OPEC as a player in the energy business.

Energy security and the future

Demand for fossil fuels

It is likely that in the next few decades fossil fuels will continue to supply most of the world's energy needs. There are a number of uncertainties surrounding fossil fuel demand.

- **Economic growth** is related to energy demand. Demand was strong between 2002 and 2006 as the world enjoyed an economic boom. Eventually oil demand outstripped supply, pushing a barrel of oil to over US$100 by summer 2008. Oil prices collapsed to US$35 in 2009 but by 2011 were again over US$100. The IMF still expects the world economy to double in size by 2040, which would clearly increase energy demand.
- **Efficiency** might lower demand. High energy prices in 2006–08 in the UK and Europe encouraged a switch to fuel-efficient cars, public transport and home insulation. If governments led a meaningful efficiency drive, the savings could be very large.
- **Population growth** is uncertain. The UN estimates 8.5 billion people by 2040 — 2 billion more than in 2005. How developed these extra 2 billion will be largely determines which energy sources they would use. Development in the '100 million+ club' (India, Brazil, China, India, Bangladesh, Nigeria, Pakistan, Mexico) would increase demand.
- **Renewable and recyclable resources**, if used more extensively, would reduce demand for fossil fuels.

Examiner tip

You need to be able to discuss the pros and cons of nuclear power in detail, including the impact of events such as the 2011 Fukushima disaster in Japan on public perception of nuclear energy.

A large question mark hangs over the future of nuclear power. A single reactor produces 1,100 MW of power, equivalent to 600 large wind turbines. However, of the 439 nuclear reactors operating in 2008, only 34 were constructed in the period 1998–2008, whereas 213 are between 21 and 30 years old. Unless a huge programme of nuclear capacity building is launched soon, nuclear power may decline in importance.

The 'peak oil' problem

Demand is less of a concern than supply. Oil and gas are flexible, 'clean' fuels compared to coal. Coal reserves will last for 150–200 years at current use rates, but coal is less energy dense than oil or gas, more costly to transport and dirtier — especially in terms of acid rain causing sulphur dioxide and nitrogen oxides.

There are concerns that oil and gas supplies will 'peak' in the near future. The production peak point is more important than when oil and gas will 'run out' because after the peak supplies will shrink and prices will rise.

The timing of peak oil is hotly disputed.

- In 2008 the Association for the Study of Peak Oil and Gas put the date at 2010.
- In 2007 the German Energy Watch Group claimed the peak was reached in 2006.
- In 2008 the UK Industry Taskforce on Peak Oil and Energy Security stated it would be reached by 2013.
- In 2006 the IMF predicted oil production rising to nearly 120 million barrels per day by 2030.

Another view is that since 2004, oil production has been 'stuck' at 80–85 million barrels per day, producing 'plateau oil' rather than a distinct peak.

Some in the oil industry argue that the huge price rises of 2007–08 were evidence of the 'peak'. Others countered with the view that supply was being limited by a lack of oil refinery capacity. There is less concern about peak gas. Most estimates suggest this is further off, perhaps occurring between 2025 and 2030. Increased exploitation of shale gas (also called tight gas) could dramatically increase the availability and longevity of natural gas supplies.

Knowledge check 8

Explain why peak production of oil is more important than the date at which oil actually runs out.

Examiner tip

You could practise drawing a peak oil sketch graph, showing both peak and plateau scenarios. Simple diagrams like this can be useful in the exam.

Business as usual?

The timing of peak oil is important. Those who believe it is close, argue that urgent action is needed to develop alternative energy sources. Even if global peak oil is years away, individual country peaks are not. UK oil production peaked in 1999 and the USA in 1970. Increasingly countries will have to rely on a shrinking number of nations, which have not peaked. These will be concentrated in unstable regions such as the Middle East (Iran, Iraq) and Africa (Nigeria, Sudan, Libya). In a future world, with tightening oil and gas supplies, what are our options? Table 6 explores three scenarios.

Examiner tip

During revision, draw up a large table to assess the environmental, social and economic advantages and disadvantages of nuclear power, biofuels, wind power and HEP.

Table 6 Energy futures?

Scenario	Business as usual scramble	New atomic age	Renewable renaissance
Energy sources	Coal, oil and gas	Nuclear power	Wind, solar, wave, hydrogen, biomass
Up to 2020	• Scramble for remaining reserves in Africa • Increased pressure to develop Arctic and Antarctic	• Race to build new reactors which are initially very costly • Political difficulties due to public perception	• Biomass/biofuels used as an 'easy option' pushing up food prices • Wind power lacks public's support
2020 and beyond	• High energy prices • Increased use of non-conventional oil • Widespread use of coal and coal gasification due to resource nationalism	• Developing world largely excluded on grounds of cost and technology • Long-term solution to nuclear waste a major challenge	• More stable supply following initial problems • Technology makes solar power very viable • Developing world begins to benefit
Environmental and other issues	• Limited progress on global warming • Ecosystem destruction • Acid rain and urban air pollution • Uses existing technology and no need to adapt to new energy sources	• Some impacts from mining uranium • Waste disposal concerns • Lower carbon dioxide emissions • Transport technology has to be redeveloped to be electrically powered	• Major reductions in carbon dioxide emissions • Large areas of land used, difficult for small nations • Requires efficiency and hydrogen production (cars) to be truly feasible

Knowledge check 9

Suggest reasons why public perceptions of nuclear power are generally unfavourable.

Rising tensions

The future may well see a rise in resource nationalism. If energy insecurity increases, countries are likely to turn to their own energy resources. In fact, this has already occurred with biodiesel and bio-ethanol in the USA. Faced with rising oil prices and lack of supply security, the USA:

- passed the Energy Policy Act 2005, with a target of 7.5 billion gallons of biofuels by 2012 and 10% of petrol must be ethanol by 2009
- passed the Energy Independence and Security Act 2007 stating that 36 billion gallons of petrol must be biofuel based by 2022

These Acts created a market for biofuels, which in turn led to huge areas of maize and soybean being grown for biofuels (Figure 9).

Examiner tip

Be clear about what biofuels are used for. They are mostly a replacement for petrol and diesel refined from fossil fuels.

The promotion of biofuels in the USA and Europe meant that by 2007–08 large areas of land that once grew food were now growing fuel. This contributed to rising global food prices and food riots in Mexico, India, Yemen, Bangladesh and Indonesia. The energy policy of one country can thus have negative consequences for others.

An alternative approach is to get hold of someone else's energy resources. This has been China's approach in Africa. China has signed bilateral agreements with African countries to ensure an oil supply. Chinese demand is set to rise from 3.5 million barrels per day in 2006 to 13 million by 2030.

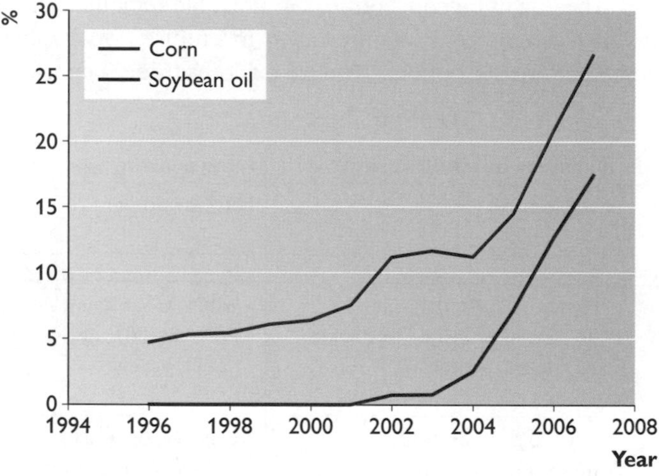

Figure 9 Share of US maize and soybean oil production converted into biofuels

There are concerns over China's move into Africa:

- bilateral oil trade agreements work against the idea of an open oil market
- China has invested heavily in (some would say meddled in) war-torn Sudan, putting its oil interests ahead of humanitarian concerns
- suggestions of links to undemocratic or despotic regimes such as Zimbabwe
- Africa's oil should help African development, not Chinese development

Knowledge check 10

Explain China's interest in Africa's energy resources.

There are some positives. Chinese investment in oil does bring money to the world's poorest continent. Chinese investment has been dwarfed by the investment of supermajors in Africa, but it is growing fast.

The alternatives

Ours is a world with large energy needs:

- the need to have continuing secure supplies in the developed world
- meeting the needs of the emerging economies
- providing energy to the least developed world, to meet basic needs

These needs may seem enough on their own, but we also need to reduce urban air pollution, protect biodiversity and tackle global warming. Addressing these environmental concerns and meeting energy demand may seem like an impossible circle to square. Some difficult choices will need to be made to ensure energy supplies and protect the environment. Some possible decisions are shown in the five energy 'R's in Table 7.

Long term, a switch to renewable sources looks inevitable. The question remains as to whether countries can make the switch in a relatively painless way, or whether players will cling to a fossil fuel model until the resources themselves begin to become prohibitively expensive. Under that scenario, conflict over resources and rising tensions are likely to be the result.

Examiner tip
Consider which of the 'R's in Table 7 are easiest to achieve and which would have the greatest impact on the energy security of a country like the UK.

Table 7 The five energy 'R's

'R'	Example
1 Refuse polluting energy sources	Strong argument for leaving tar sands and oil shales in the ground due to environmental costs of extraction, and the need to use energy (natural gas) in processing
2 Reduce overall consumption	Energy efficiency standards for vehicles and domestic appliances. Use tax system (green taxation) and polluter pays principle; 'carbon credit cards' to encourage personal efficiency
3 Research more sustainable and affordable technologies	Wind, solar photovoltaic, passive solar, biomass and micro-hydro developed as low cost, portable or micro-installation systems for the developing world or as domestic systems
4 Recycle waste, and convert into useful energy	Landfill — methane gas can be siphoned off and used to generate electricity, and municipal waste can be converted to useful energy in Combined Heat and Power (CHP) plants
5 Replace inefficient/ wasteful technologies with improved ones	Phase out incandescent lights and replace with CFL; replace petrol and diesel with hydrogen or electricity produced from renewable sources

Synoptic links

The links below show how **energy security** links to the three synoptic themes (players, actions and futures), and to other units you have studied as well as to wider global issues.

Players

Large companies, either as TNC 'supermajors' or state-owned companies, play a key role in energy supply including exploration, extraction, processing and delivery. These companies are some of the most powerful economic entities in the world. Some are beginning to invest in renewable technology.

Governments influence a nation's energy mix. Many are beginning to promote renewable resources as a way of increasing energy security.

Pressure groups try to influence energy policy, and press for different energy sources, or argue against their construction, e.g. wind farms and nuclear power stations.

Actions

Because energy security is so important, it is often not left to the **market**. Governments are frequently heavily involved in energy using targets, quotas, subsidies or the legal system. There is a **global** market in energy, but energy systems tend to be in the form of '**national grids**'. Critics argue energy policy would be more sustainable if it was more **locally** controlled.

Futures

A **business as usual** energy future would involve continued dependence on fossil fuels, with environmental and supply implications.

A more **sustainable** energy future could mean switching to renewable sources over a period of decades, possibly including large-scale use of nuclear power.

A more **radical** approach might involve drastic cuts in personal energy use, local/household renewable supplies and a rapid decrease in fossil fuel use forced by very high taxes on polluters.

Links to other units

Unit 1

- **World at risk**: the causes of global warming.

Unit 3

- **The technological fix?** contrasting energy technologies, their impacts and availability.
- **Superpower geographies**: the increasing power of Russia and the Gulf States, based on their large remaining fossil fuel reserves.
- **Bridging the development gap**: the role of China in exploiting Africa's oil and energy development projects such as the Three Gorges Dam.

Knowledge check 11

Are energy resources alone enough to make a country a superpower?

Unit 4

- **Pollution and human health**: the impacts of pollution from power stations on people and their environment.

Links to wider global issues

- **Global warming** is a key wider issue linked to energy security. In order to meet obligations under the 1997 Kyoto Protocol to reduce carbon dioxide emissions, governments have had to rethink energy policy. Many European governments have shifted towards renewable sources. The EU Emissions Trading Scheme penalises heavy polluters but rewards those who increase energy efficiency and reduce emissions.

- The **development gap** can be illustrated with reference to energy. Energy use in the least developed countries is very low per person, and people rely on recyclable and renewable resources such as fuel wood, dung and biogas. NICs are often the heaviest users of coal, whereas the developed world uses cleaner fuels such as gas, and increasingly, technologically advanced wind and solar renewable sources.

Summary

- Non-renewable fossil fuels currently provide 85% of global energy supply.
- Alternatives to fossil fuels include recyclables such as nuclear power (re-use of spent fuel rods), biofuels, and renewables such as wind, solar etc.
- These primary resources are used to generate secondary energy such as electricity. So, globally, nuclear power provides only 6% of the world's energy, but around 18% of the world's electricity.
- Demand for energy is closely linked to level of economic development, whereas supply is largely related to physical factors.
- The imbalance between supply and demand leads to some energy sources such as oil being used as geopolitical weapons by energy rich suppliers such as Russia and the Middle East.
- As energy is so vital, all countires strive to achieve energy security, i.e. ample supplies of affordable, reliable resources with minimum damage to the environment.
- Each country has a distinctive energy mix based on physical, socioeconomic and political factors.
- Key players in the energy game include governments, energy TNCs, cartels such as OPEC, as well as environmental groups and consumers.
- In the foreseeable future, they are major drivers towards a future based on reduced reliance on fossil fuels, rising demands from BRICs, and concerns over peak oil and rising GHGs (especially from coal) leading to global warming.
- The development of unconventional oil and gas supplies, and clean coal technology, are high-cost energy options.
- The changeover may be very gradual, largely because of concern over the safety of nuclear power, the dubious environmental credentials of biofuels, and the high cost and lack of consistency of supply of many renewables.
- The development of a sustainable future, with greater efficiency of use and conservation, will play a vital role as many countries could experience an energy gap.

Water conflicts

Water, like energy and food, is a fundamental need. The problem is that it is predicted there will be a **world water gap** between growing demands and diminishing supplies. Population growth (possibly an additional 3 billion people by 2025), economic development (especially in RICs and NICs) and rising standards of living, all increase demand. If we follow a business as usual approach, there will be a 56% increase in demand if all water uses (agriculture, industry, domestic, HEP) are combined. Even with more sustainable use there is still an estimated 20% increase in demand predicted. This will lead to conflicts between the various users.

Water supplies are very unevenly spread across the world. Two thirds of the world's population lives in areas receiving only 25% of the world's annual rainfall. This is a second potential source of conflict between countries and regions with a water deficit and those with surpluses, especially when they share a very large river basin.

Knowledge check 12

Explain why the widening water availability gap reflects the widening development gap.

A third underlying conflict arises from the **water availability gap** between the 'have nots' largely in developing nations (especially sub-Saharan Africa) and 'haves' in developed nations. This gap is widening and reflects the development gap. There is also an imbalance in usage with rich countries using up to ten times more water per capita.

Other points to note include:

- Climate change will affect many developing nations which lack the resilience and technology to adapt (some sources such as glaciers in the Andes, Kenya highlands and Himalayas are disappearing).
- In coastal areas, over-abstraction and rising sea levels are leading to salt water contamination, e.g. in South Sea Islands.
- The costs of a safe water supply in developing megacities are rising. Already slum users pay up to six times as much as UK users as they rely on deliveries of bottled water from vendors.
- There is widespread pollution of supplies, especially in developing countries (1.5 billion people without safe water, 2.3 billion lack adequate sanitation).
- Developing areas are experiencing rapid population growth — 2–3% per year — and therefore rising demand.

The geography of water supply

Global water supplies are linked to three main physical factors:

- **Climate** determines the global distribution of water by means of annual and seasonal rainfall distribution, or snowfall.
- **Rivers** transfer surface water across continents.
- **Geology** controls the distribution of aquifers, which supply underground water.

The rivers and aquifers are collectively called **blue water flow** — the visible part of the hydrological system.

Figure 10 shows the availability of the world's water supplies. Only 1% of the Earth's total water resources are easily accessible as surface water.

Examiner tip

These pie charts require analysis. Write a paragraph in which you explain why there is such limited availability of water for human use.

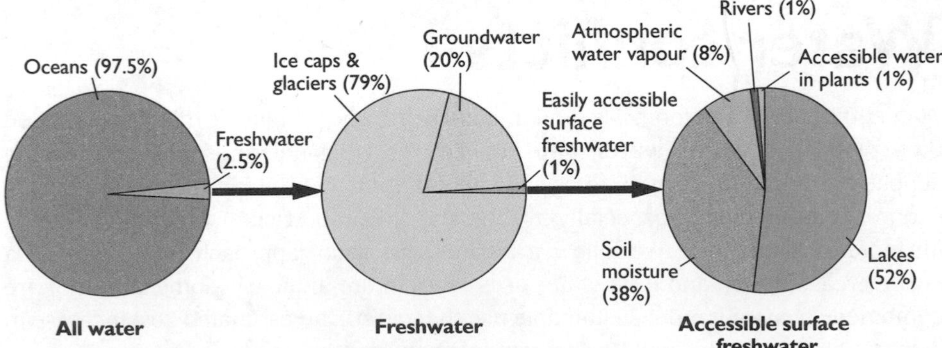

Figure 10 The availability of the world's water

Water stress, scarcity and vulnerability

Globally the picture is of increasing risk for large numbers of people.

- By 2025 it is estimated that nearly half of the world's population will be water **vulnerable** (under 2,500 m^3 per person per year).
- Many of these will experience **water stress** (under 1,700 m^3 per person per year) especially in the Middle East, North Africa, sub-Saharan Africa and southwest USA.
- **Scarcity** occurs when the annual supply of water per person drops below 1,000 m^3.
- There are two types of water scarcity (as shown in Figure 11):
  - **physical** scarcity occurs when more than 75% of a country's or region's river flows are being used (25% of the world's population lives in such areas)
  - **economic** scarcity occurs when the development of blue water sources is limited by lack of capital and technology

Many parts of the continent of Africa — some 1 billion people — currently have physical availability but only access 25% of the water supplies because of the high levels of poverty in these LDCs.

With the onset of climate change and the associated deterioration of ecosystems, combined with rapid economic development in Asia (as a result of the emergent superpowers of China and India), the number of people experiencing water stress is expected to rise. By 2050 there will be 4 billion people experiencing stress and a further 1.5 billion — especially in the Middle East and parts of the continents of Africa and Asia — experiencing water scarcity.

Knowledge check 13

Explain why south and east Asia have been identified as areas of increasing water stress.

There are some regions within developed countries, such as the southwest USA and parts of Mediterranean France and Spain, that are also areas of concern and shortages may need to be solved by water transfers or widespread desalination. Even southeast England is potentially vulnerable to serious water shortage.

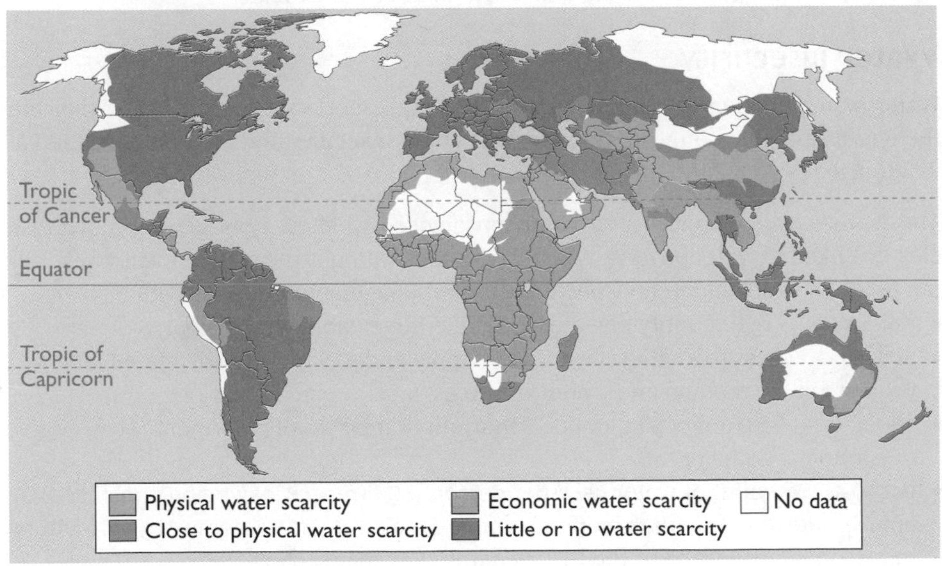

Figure 11 Global distribution of water scarcity, 2006

Examiner tip

Describing maps is a high-level skill. Using an atlas, describe the distribution of the **four** categories of water scarcity shown.

Human impacts on water availability

Quantity

Humans can remove water from rivers and groundwater sources, whether for drinking, for irrigation or for industrial purposes. If they **over-abstract**, supplies cannot be replenished in time and reserves will be lost because rainfall can never fully recharge the underground stores. Unintended consequences of over-abstraction (groundwater mining) can lead to subsidence as in Mexico City, salination of wells and boreholes, and loss of valuable wetlands by drainage or as a result of incursions by the sea.

Quality

Human actions can pollute both surface water and groundwater supplies.

- Sewage disposal in developing countries causes water-borne diseases such as typhoid, cholera and hepatitis. As many people are forced to use unsafe water, it is estimated that up to 135 million people may die from water-borne diseases by 2020 (WHO). Poor water leads to poor health. In Bangladesh millions of deep tube wells were sunk to provide pure drinking water free from sewage contamination, which was killing up to 200,000 children a year. The new water supply proved to be contaminated with naturally occurring arsenic, which is slowly poisoning the entire population in a tragedy greater than Bhopal or Chernobyl.
- Chemical fertilisers used by farmers contaminate groundwater as well as rivers, and eutrophicate lakes and rivers, which leads to hypoxia and the formation of dead zones in seas.
- Industrial waste is dumped into rivers and oceans. Heavy metals and chemical waste (PCBs) are particularly toxic.
- Sediments trapped behind big dams can damage fish stocks.

Examiner tip
Research a large river such as the Ganges or Nile and draw a sketch map to identify the key sources of pollution.

Water insecurity

Water insecurity means not having access to sufficient, safe (clean) water. Typically the world's poorest countries are the most water insecure, and are also classified as water scarce.

The water poverty index (WPI) has been devised to show how general poverty is closely linked to water poverty. A country's index rating is the sum of five scores each out of 20, in which different aspects of water management are assessed:

- resources — the quantity of water available (renewable water supply)
- access — access to an improved water supply and to sanitation; irrigated land as a proportion of cropland and water resources
- capacity — GDP per capita, under-5 mortality rate, school enrolment rates, degree of economic equality
- use — the amount of water used per person (50 litres per day being considered an appropriate amount); the amount of water used for a sector (domestic, agriculture, industry) in proportion to the GDP generated by that sector
- environmental impact — water quality and stress, and the importance attached to water and environment

Knowledge check 14
Explain why in general developing countries have less favourable WPIs.

Wealthy countries such as Canada and Sweden score 80/100, whereas LDCs such as Burkina Faso and Ethiopia score under 40/100. However, note that marked differences can be seen in the profiles of countries with similar index ratings. For example, the USA (65/100) scores highly on access but poorly on use, while Equatorial Guinea (68/100) scores highly on use but poorly on environmental impact. Also, a poor country may have a higher score than a rich country in one aspect, especially use (e.g. USA vs Sudan), despite having a much lower overall score.

The risks of water insecurity

Water supply problems

Water has been described as the 'lubricant of development'. Secure water supplies are essential to economic development.

- Water supplies are essential to support irrigation (some 70% of total use) so that food production can be increased. High-tech cash crop farming typified by the Green Revolution poses very high demands on water.
- Water is needed to support industries — even high-tech companies use huge quantities of water for manufacturing silicon chips or making beverages (e.g. in Kerala the Coca-Cola bottling plant uses 4.5 million litres a day taking all the local water).
- Water supplies are needed for hydroelectric power production and also for cooling in thermal power processing.
- Safe, secure water supplies ensure better health and higher standards of human wellbeing, hence the provision of such supplies was listed as one of the eight Millennium Development Goals.

Examiner tip

Research three common waterborne diseases and describe their impacts.

However, the era of cheap and easy access to water is ending, which poses a serious threat to economic development. Water shortages are considered an even greater threat than fossil fuel shortages as there is no substitute for water.

The extraction and use of water resources can lead to environmental damage and supply problems for both groundwater and surface water supplies. A useful case study is that of the former Soviet Union: 75% of the surface water and 40% of the groundwater was polluted. The Aral Sea (formerly the size of Ireland) declined to 10% of its original size as a result of the diversion of rivers such as the Amu Darya and Syr Darya to provide irrigated water for cash cropping of cotton. The salination of the Aral Sea is an ecological and environmental catastrophe. Recently there has been a World Bank-funded programme to restore the northern part of the Aral Sea. Following the break up of the Soviet Union, there is potential for conflicts between the newly independent states.

Water conflicts

When demand for water overtakes available supply and several stakeholders (players) wish to use the same resource, there is a potential for conflict. Competing demands for water for irrigation, energy, industry, domestic use, recreation and conservation can lead to tension within a country and between countries. Figure 12 summarises the components of these sites of potential conflicts, known as water hotspots, pressure points or pinch points.

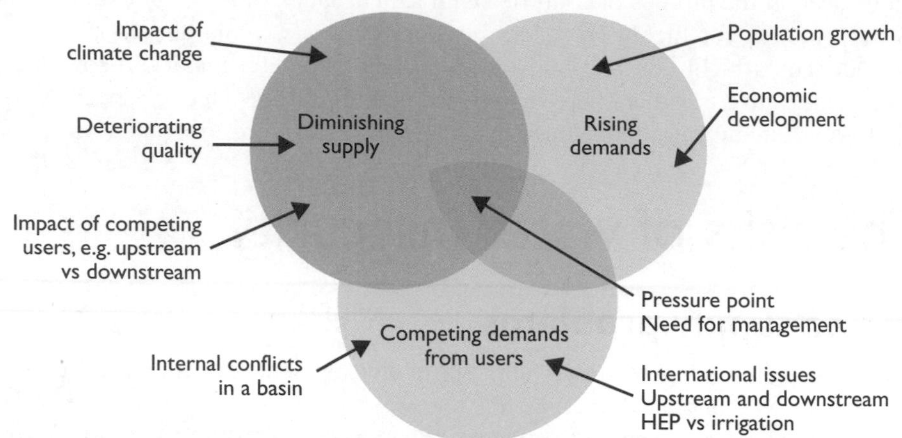

Figure 12 Water pressure points

However, between 1950 and 2000, out of nearly 2,000 international events, only 25% led to any form of conflict. Of these conflicts nearly two thirds were about the quantity of water available, especially where upstream users had diverted and used water in a river basin at the expense of lower basin users. The other main source of conflict was over the impact of dams (25%). Over 65% of the world's large river basins are affected by dams and diversions. Cooperation as a result of positive actions between river users involved a wide range of issues. For example, the Mekong River Committee was established between Thailand, Laos, Cambodia and Vietnam and dealt with HEP, flood control, infrastructure, technical cooperation etc.

Examiner tip
When writing about water wars you need to support your work with a **range** of examples. Select four contrasting examples from Table 8 to research the reasons for the conflicts.

An important question is: will conflicts become more dominant in the twenty-first century as water insecurity increases? The likely sites of conflicts are where countries are habitually in conflict such as Israel and Palestine, or India and Pakistan/Bangladesh.

Table 8 lists the main current pressure points of surface water use.

Table 8 Main water pressure points

Location	Reason for pressure point
Tigris–Euphrates Basin	Concerns from Iraq and Syria that Turkey's GAP project will divert much of the water via a series of irrigation dams. Syria has also developed dams, which in the 1990s initially led to conflict with Iraq
Jordan River	Use of the Jordan, largely by Israel but also Syria, Lebanon and Jordan, has reduced the flow of the river to a mere trickle. This also affects supplies to Palestine's West Bank
Ganges–Brahmaputra	India has built dams such as the Faraka, which has reduced the flow of the river into Bangladesh
Syr Darya and Amu Darya, Central Asia	Turkmenistan, Uzbekistan and Kazakhstan need more summer water for irrigation, but water has been diverted by schemes in Tajikistan/Kyrgyzstan
Colorado Basin	States in USA dispute their allocation of water from the Colorado, which is so great that the quantity and quality reaching Mexico does not reach the standard agreed
Nile Basin	While agreements exist, schemes developed in Ethiopia and Sudan may threaten supplies to Egypt

Groundwater conflicts occur in similar areas to surface water ones — for example between Israel and Palestine over the use of the mountain aquifers.

Pressure points include the Middle East, Arabia and North Africa where water has been so over-abstracted that it cannot be replenished, and also in southwest USA, North Africa, Spain and the Upper Indus where water is currently abstracted at a faster rate than it can be replenished.

Many of the subterranean aquifers straddle international boundaries. The issues of **shared groundwater usage** are highly complex.

- Supplies are underground so it is difficult to understand the problem as it takes years for an effect to show.
- It is difficult to negotiate an equitable and reasonable share for each nation to exploit, as often the extent is not mapped and nobody knows who owns what.
- More powerful developed nations 'mine' the groundwater more efficiently because they have deeper wells and more efficient pumps at the expense of their neighbours.
- Legislation by the UN to sort out water sharing of aquifers between nations is only just being written.

Knowledge check 15

Explain why groundwater conflicts tend to be more difficult to resolve than disputes over surface water.

Nevertheless, as with surface water, conflicts are not inevitable. In North Africa, Chad, Egypt, Libya and Sudan have negotiated an agreement to share water in the Nubian Sandstone, and Nigeria, Niger and Mali are currently negotiating for joint management of Lullemedan aquifers in the Sahel Zone.

Water geopolitics

As countries compete for increasingly scarce water resources, international agreements and treaties have been drawn up as a framework to manage the distribution and use of shared water supplies.

Under the **Helsinki Rules** there is general agreement that international treaties must include concepts such as 'equitable use' within a drainage basin when devising criteria for **water sharing**. Criteria include:

- natural factors
- downstream impacts
- social and economic needs
- prior use

The problem is that upstream nations assert their rights to territorial sovereignty whereas downstream nations claim territorial integrity (right to receive the same amounts of water as in the past). The player with the greatest military, economic and political power is the winner — so legislation is vital. When a multilateral aid scheme is granted by the World Bank, water sharing agreements must be 'built in' if it involves a transnational river.

Knowledge check 16

Identify the national and international conflicts linked to the Colorado Basin.

Laws are passed such as the 'Law of the River' which operates to share out the waters of the Colorado between the US states and Mexico. Even so there are frequent disagreements — especially when supplies are scarcer than usual because of a period of drought.

Water transfers

Water transfers involve the diversion of water from one drainage basin to another either by diverting the river itself or by constructing a large canal to carry available water from the area of surplus to the area of deficit.

Examiner tip
Search on the internet for new technologies being used in the desalination process, e.g. nano-technology, which are cutting costs, and decreasing the environmental damage of desalination plants.

Currently there are many large-scale water transfer schemes in operation. The engineering itself and the actual water transfers have been successful but there are many environmental and social disadvantages, as shown in Figure 13.

Table 9 shows some of the existing water transfer schemes and some of the proposed schemes. These increasingly massive engineering schemes may be a techno-fix for water redistribution, but they have potentially huge environmental costs. The economic costs are so high that in countries such as Spain and Israel desalination is increasingly viable.

Knowledge check 17
Explain what is meant by eutrophication.

Examiner tip
Use the internet to look at the latest state of play on transfer schemes. Some will go ahead — e.g. China and Israel — others may never be more than plans.

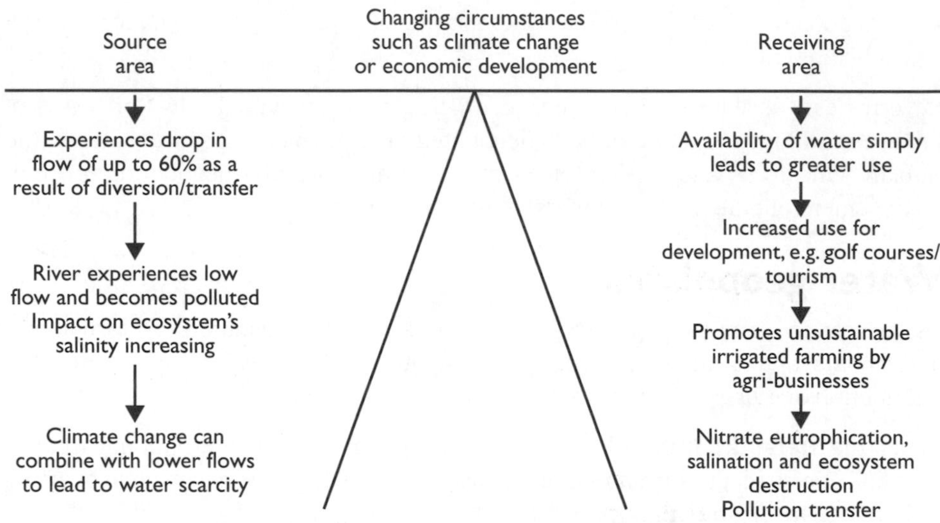

Figure 13 Water transfer issues

Table 9 Examples of water transfer schemes: existing and proposed

Existing schemes	Details
Tagus–Murcia transfer in Spain	Takes water from Tagus River by canal to the drought stricken area of Almeria–Murcia–Alicante to provide water for 700,000 new villas and the tourist industry and irrigating cash cropping areas
Snowy Mountains scheme in southeast Australia	Water is transferred from the storage lake of Eucumbene westwards by the Snowy–Geehi tunnel to the head waters of the Murray to irrigate farms and provide water to an increasingly drought stricken area
Melamchi project in Nepal	Water is diverted from the Melamchi river via a 26 km tunnel to water stressed areas in the Kathmandu basin. In return the residents of Melamchi are provided with improved health and education services
Proposed schemes	**Details**
South to North transfer project in China	Began in 2003 and will take 50 years to complete, costing up to US$100 billion. It will transfer 44.8 billion m^3 of water per year from the relatively water-secure south to the drought stricken north via 1300 km of canals linking the Yangtze to the Yellow Huai and Hai rivers

Proposed schemes	Details
Ebro Scheme, Spain	Following on from the Tagus scheme, 828 km of canals will be built to divert the waters of the Ebro to southern Spain
Israel – transfer from any neighbours who would agree	Israel has a huge water deficit and plans for several schemes including transferring waters from the Red Sea to top up the Dead Sea. Another scheme to tap surplus waters from the Mangarat River in Turkey, possibly by water tanker or undersea pipeline (subject to terrorism), has since been shelved
Projected water transfer systems in Russia	Russia plans a whole series of schemes diverting rivers such as the Ob to the drought stricken area of the Aral Sea. The diversions could have major implications for the Arctic Ocean, as they would affect salinity
Projected water transfer systems in India	India plans to develop a national water network, to ensure a better distribution of supplies to water deficit areas such as the Deccan plateau A similar water grid has long been considered for the UK so that the surplus from the northwest could be moved to the southeast
Projected transfer projects in North America	Canada is a water surplus country. NAWAPA is a scheme to take water from Alaska and northwest Canada to southern California and Mexico. A further scheme — Grand Canal — could take water from Hudson Bay to the Great Lakes

Water conflicts and the future

Figure 14 shows that total projected water withdrawals are predicted to reach over 5,000 km^3 per year. This is likely to have a considerable impact on the natural environment as well as on people's health and wellbeing. There will be a knock-on effect to the food security of the world's poorest people.

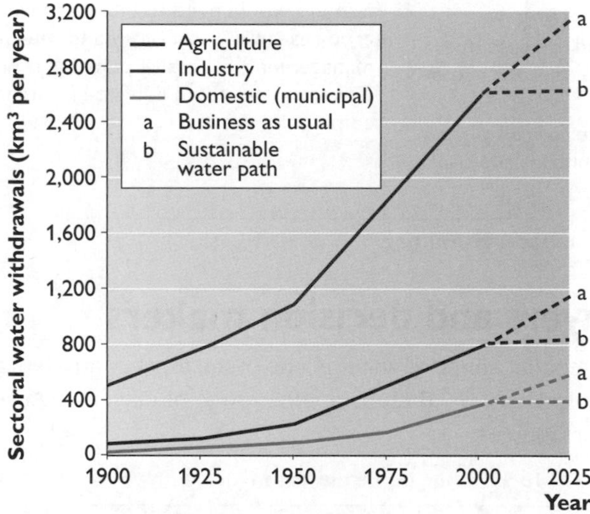

Figure 14 Rising trends in water use

In 2002, the International Food Policy and Research Institute used a computer model to examine the implications of three alternative futures for global water (and food) supply and demand. Calculated for 2025, these futures are shown in Table 10. Clearly the business as usual scenario will be unsustainable in the long term. The most worrying scenario is that of water crisis, which shows how mismanagement of water resources or climate change could threaten our water and food supplies and lead to wider geographical problems, including conflict. Alarmingly, some features of this scenario may be starting to occur already. Note that all the predictions are tentative as they involve uncertainties such as the scale, severity and impact of climate change. Recently the concept of a virtual water footprint has been included in an assessment of eco-footprints. With so much imported food, the UK has a very marked virtual water footprint.

Knowledge check 18

Which scenario will be best for food security and why?

Table 10 Alternative scenarios for water by 2025

	Water changes by 2025	Wider impacts
Business as usual	• Water scarcity will reduce food production • Consumption of water will rise by over 50% • Household water use will increase by 70% (mostly in developing countries) • Industrial water demand will increase in developing countries	Developing countries will become reliant on food imports and experience increased hunger and malnutrition. In sub-Saharan Africa, grain imports will more than triple. In parts of the western USA, China, India, Egypt and North Africa, users will pump water faster than aquifers can recharge
Water crisis	• Global water consumption would increase further, mostly going to irrigation • Worldwide demand for domestic water would fall • Demand for industrial water would increase by 33% over business as usual levels, yet industrial output would remain the same	Food production would decline and food prices, especially of cereals, would increase rapidly. In developing countries malnutrition and food insecurity would increase. Dam building would decline because of fewer potential sites and key aquifers in China, India and North Africa would fail. Conflict over water between and within countries would increase
Sustainable water	• Global water consumption and industrial water use would have to fall considerably • Environmental flows could be increased dramatically compared to other scenarios • Global rain-fed crop yields could increase due to improvements in water harvesting and use of sustainable farming techniques • Agricultural and household water prices might have to double in developed countries and triple in the developing world	Food production could increase slightly and shifts occur in where it is grown. Prices could fall slowly. Governments, international donors and farmers would need to increase investment in crop research, technology and reforms in water management. Excessive pumping is unsustainable. Governments could delegate farm management to community groups

Source: **www.worldwatervision.org**

Water players and decision makers

The use and subsequent abuse of water is one of the most controversial issues facing the world. As Table 11 shows, there is a wide range of players involved in any issue relating to water resources.

Table 11 Some of the players involved in water issues

Category	Players
Political	International organisations (e.g. UN) responsible for MDGs; government departments (e.g. DEFRA); regional and local councils; lobbyists and pressure groups who form to fight particular issues such as the building of mega dams
Economic (business)	World Bank and IMF fund mega projects and ensure legislation is in place for transboundary schemes. Developers of mega schemes. Transnational water companies (utilities) that run the supply business. TNCs and businesses that are large users (agriculture, industry, energy and recreation)

Category	Players
Social (human welfare)	Individuals, residents, consumers, land owners, farmers, who feel access to water is a human right. Health officials who try to ensure safe water. NGOs such as Water Aid or Practical Action that develop sustainable schemes for the poor in LDCs
Environmental (sustainable development)	Conservationists who fight hard engineering schemes or seek to save wetlands. Scientists and planners who develop new schemes. BINGOs such as WWF try to influence World Water Policy. UNESCO/FAO/IUCN that operate globally

Controversy 1

Social players see access to clean, safe water as a human right whereas political players see water as a human need, which, like food and shelter, can best be provided through market mechanisms. It is estimated that in order to meet the UN's Millennium Goal of halving the proportion of population without access to improved water supply and sanitation it would cost around US$200 billion. This is to supply pipes, sewage systems, water treatment plants etc. Governments, especially in low income countries, cannot afford these huge sums so rely on privatised profit-led organisations (transnational utility companies) to develop water supply infrastructure. Clearly these companies need to be offered favourable terms to invest, which can lead to rising prices for the consumers, and minimal investment in environmental improvement.

Examiner tip

Use your textbooks to research case studies so that you can identify the different values and positions of the players involved. Two controversies are outlined here. Also prepare for an assessment of the **relative** importance of named players.

Controversy 2

In order to keep pace with rising demand, political and business players favour hard engineering schemes such as mega dams, water transfer projects and clusters of desalination plants. Inevitably, many of these schemes have very high social, economic and environmental costs and are opposed by social and environmental players who favour more sustainable approaches.

Responses to rising demands

Managing future water supplies will require action at a variety of levels, ranging from large-scale projects funded by governments and agencies such as the World Bank, down to changing individual consumers' attitudes to water use at a local level. Possible actions include those outlined below.

Hard engineering projects

- Currently there are 845,000 dams in the world of which some 5,000 could be considered to be mega dams. Two thirds of all surface freshwater is obstructed. The issue is that most of the best sites have been used, and large schemes such as Three Gorges Dam are considered unsustainable. Over half of dams are primarily used for irrigated agriculture.
- A similar trend can be noted with large-scale water transfers whereby the schemes may have huge environmental consequences for source and receiving areas.
- Desalination is growing at an exponential rate, but is concentrated in water stressed, technologically advanced countries such as Saudi Arabia, UAE, Kuwait, USA, Spain and Japan. Improved technology (osmosis membrane) filters salt from brackish water, which leads to lower costs, but filtering sea water is still very costly.

Examiner tip

Be very careful to identify the prime purpose of any mega dams chosen for case studies. Avoid describing the story of the dams.

Water conservation

Developing sustainable strategies to conserve water supplies is vital in the attempt to combat climate change.

- In agriculture the maxim has to be 'more crop per drop'. Sprinkler and surface flood irrigation is steadily being replaced by modern spray technology and more advanced drip irrigation, which uses less water.
- Conservation of industrial water is about reuse (**grey water**). Waste water can frequently be recycled using filters and chemicals. In Singapore, where it is called 'new water', it plays a major role in water conservation.
- For **domestic users** the compulsory provision of water meters makes consumers more careful about water use. There are a number of strategies they can use in the kitchen and bathroom to reduce water use (e.g. using an eco kettle, a low flush toilet). In the garden, **water harvesting** via rain butts is an important strategy.
- Water companies carry out projects to cut down on leakage from broken pipes.
- There are a number of sustainable water projects that can be developed in low-income countries to provide villages with accessible clean water and also to conserve water (e.g. magic stones — see **www.wateraid.com**).
- **Restoring damaged wetlands** to their natural state can renew vital water stores; the restoration of the northern part of the Aral Sea is currently the world's largest restoration project.

Knowledge check 19

With reference to examples of schemes for improving water security, distinguish between intermediate and high-technology.

Integrated water management

This is now generally regarded as essential to dealing with water scarcity — see Figure 15.

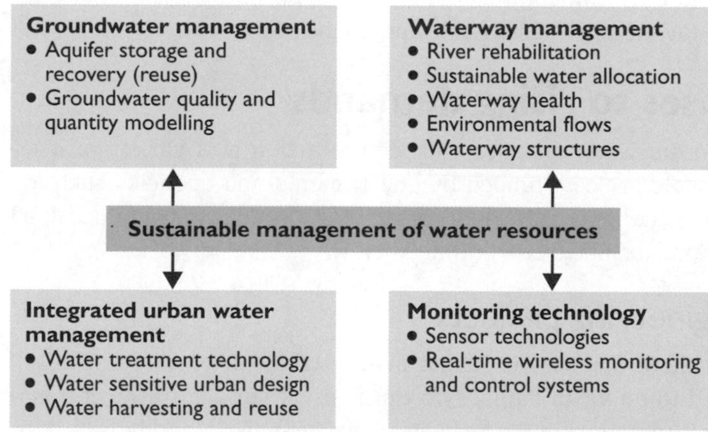

Figure 15 Integrated sustainable water resource management

The physical resource is the starting point. Satellite images and water accounting are used to determine how much water there is and how productively it is used. Also important is the way the water is managed within a basin community and its impact on the surrounding environment. Truly integrated management is a bottom-up approach designed to overcome water poverty.

Knowledge check 20

Distinguish between integrated and sustainable water management.

Synoptic links

The links below show how **water conflicts** links to the three synoptic themes (players, actions and futures), and to other units you have studied as well as to wider global issues.

Players

The supply and development of water resources involves a wide range of players. Both the development of increasingly scarce resources and supplying safe, clean water to all homes in developing countries leads to major conflicts between players. There are public/private issues, exploitation/conservation issues and supplier/user issues.

Actions

Water, unlike energy, is frequently left to market forces. The resource side is usually government managed, but the supply side is increasingly controlled by private companies.

Futures

The only future for water management is the sustainable route, which is heavily focused on conservation of existing supplies. A more radical approach may be needed if the tipping point of climate is reached.

Links to other units

Unit 1

- **World at risk**: impact of short-term climate change.

Unit 3

- **The technological fix?** role of major engineering projects and appropriate technology schemes.
- **Bridging the development gap**: the role of water as the 'lifeblood' of development.
- **Superpower geographies**: the role of mega engineering projects in raising the profile of China and India.

Unit 4

- **Life on the margins**: the vital role of finding and funding water to develop marginal areas.
- **Pollution and human health**: impact of water pollution on health.

Links to wider global issues

- **Global warming** and **climate change** are responsible for an increasing water crisis.

Knowledge check 21

Give examples of two areas where climate change will lead to future water stress.

- The **development gap** can be illustrated with reference to water — access to safe clean water in LDCs is not only more difficult but also the cost of water is far higher.
- **Sustainability** can be illustrated with respect to the use of water as a resource.

Summary

- Water is considered the lubricant of development as it is needed for agriculture to provide food security, industrial development and for domestic use.
- A world water gap is developing between rising demand from population increase and economic development and diminishing supplies.
- Resources are distributed unevenly across the globe, with 66% of the world's people in areas receiving only 25% of the rainfall.
- Water stress therefore occurs for physical reasons.
- It also occurs for economic reasons (the water availability gap).
- Nearly 25% of the world's peoples lack access to clean, safe water as the low-income countries lack capital or technology to develop their water supplies.
- The results of this water vulnerability, stress and scarcity are water pressure or pinch points. These can occur within a country (for example the Colorado River in southwest USA) or, more commonly, between countries that have to share surface water or groundwater supplies.
- Only around 5% of pinch points generate conflicts, usually in areas of extreme shortage, for example the Middle East, where countries are in dispute over a number of issues.
- The impacts of climate change are likely to exacerbate conflicts in the future as water security declines.
- Key players in the water business include governments, water TNCs, major business users (e.g. farmers), NGOs providing sustainable solutions, the UN (which develops rules for international water use) and consumers.
- A number of controversies exist between the players as to who pays the cost of providing for the poor peoples of the world, and what sort of solutions should be used.
- Hard engineering or high-tech solutions such as mega dams and reservoirs, water transfers and clusters of desalination plants, are sometimes less sustainable than smaller-scale projects using intermediate technology.
- Ultimately, integrated management and sustainable solutions are required to manage the world's diminishing water supplies.

Biodiversity under threat

Defining biodiversity

> Biological diversity (now known as biodiversity) means the variability among living organisms from all sources — terrestrial, marine and aquatic ecosystems and the ecological complexes of which they are part: this includes diversity within species (1), between species (2) and of ecosystems (3 — see Table 12).
>
> UN Convention of Biological Diversity 1992 Rio Summit

Table 12 Biological diversity

Type of diversity	Examples
(1) Genetic diversity: the diversity of genes found within a species, e.g. a type of whelk or a plant such as cereal	Genetic variability among populations or individuals can determine the degree of resistance to pests and diseases A broad gene pool is vital to combat diseases and climate change. Agro-ecosystems have been reduced by plant breeding to artificial monocultures (genetic erosion)

Type of diversity	Examples
(2) Species diversity: the variety of plant/animal species in a given area (habitat); it is a measure of species richness, i.e. the numbers of different organisms	Diversity of species bolsters ecosystem resilience to withstand threats such as climate change. Removal of a key species, such as bees, has a huge impact on the **functioning** of an ecosystem, i.e. nutrient cycling and energy flows. Endemics (unique and rare species) are especially important. Where there are few factors which limit growth, as in tropical areas, high primary productivity leads to a complex and diverse food web with many **ecological niches** leading to high biodiversity
(3) Ecosystem diversity: the variety of different ecosystems and the habitats surrounding them in a given area; it includes biotic and abiotic components	The two ecosystems with highest biodiversity are tropical rainforests and coral reefs. The high level of ecosystem diversity adds to the value of goods and services. Equally in a SSSI such as Oxwich, South Wales, there is a wide variety of habitats with the small reserve, which leads to a very high level of ecodiversity

Factors influencing biodiversity

Globally, biodiversity levels vary widely across the land and oceans. Both physical and human factors influence levels of biodiversity and these factors operate at a variety of scales from global to local (Table 13). Global **physical factors** such as variations in climate, play a major role in controlling the presence or absence of limiting factors, such as temperature, humidity, light availability and nutrient supply. An absence of limiting factors leads to high levels of primary productivity and the energy produced leads to high levels of biodiversity. Conversely, where limiting factors are strongly evident, such as cold temperatures, aridity, darkness, this will lead to low levels of biodiversity as in polar or desert regions. Another key factor is the **size** of an area, as the larger the continuous area the more species that can flourish in it. (Hence the recent conservation mantra 'size matters' and the creation of huge transnational conservation areas such as the Peace Parks of Africa.)

Locally there are numerous factors that may have an impact — usually disturbance from a natural disaster (hurricanes or coral reefs) or quasi-natural disasters such as wildfires (Victoria 2009), or even hunting, fishing, slash and burn farming, or eutrophication from high-tech agriculture.

Table 13 Factors influencing biodiversity levels

Physical factors	Human factors
• Climate, e.g. temperature, rainfall, amount of light (limiting factors) • Latitude, altitude and gradient • Vegetation and rate of nutrient recycling • Age, size and topography of area • 'Islandness' and endemism • Climate change	• Level of protection/management • Level of poverty • Direct actions exploiting flora and fauna, hunting, fishing, over-harvesting • Clearance for agriculture leading to deforestation • Growth of human population and rate of development/use of technology • Local ecosystem factors; succession disturbance; competition/colonisation; dispersion rates

When considering **human factors** it is necessary to consider both direct factors such as hunting, and indirect human factors such as climate change. It is also vital that human factors are considered as both negative (threats) and positive (spectrum of conservation strategies — see Figure 22, p. 41).

Examiner tip

Questions often ask you to look at the impact of factors at a specified scale. Check the wording of the question **very** carefully.

Examiner tip

A2 questions require you to assess. Be prepared to assess the relative importance of physical and human factors influencing biodiversity.

Hotspots

Hotspots are areas of high biodiversity. The initial terrestrial hotspots (25) designated in 1999 as a result of the work of N. Myers, covered only 1.4% of the Earth's land surface yet contained 44% of all known plant species and 35% of all known animal species. Figure 16 summarises the criteria for hotspot designation.

Knowledge check 22

Explain what is meant by endemism.

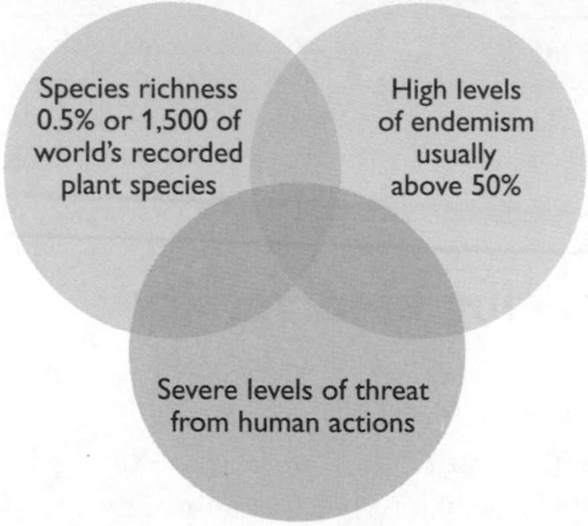

Species richness 0.5% or 1,500 of world's recorded plant species

High levels of endemism usually above 50%

Severe levels of threat from human actions

Figure 16 Criteria for hotspot designation

These hotspots are conservation priorities because of their high levels of biodiversity and endemism. But they are also under the greatest threat, which leads to environmental degradation of these valuable environments. Many people argue that with scarce eco-funds it is logical to save the 'best bits'.

Most of the original hotspots were located in **tropical** areas, especially rainforests — many in LDCs where poverty is the root cause of the threats to them. However, a criticism of this evaluation was that coverage of the world's ecosystems was uneven because marine areas and many unusual terrestrial ecosystems such as those found in polar regions were initially excluded. The development by WWF of the concept of Global 200 eco regions, which would conserve representative samples of **all** ecosystems as a priority, is an alternative approach.

Recently, Conservation International has updated the analysis of the Earth's biodiversity and has identified 34 terrestrial hotspots that between them contain over 50% of the Earth's plants. This has led to a much wider locational spread and includes several areas within the Mediterranean basin.

Knowledge check 23

Suggest reasons why marine ecosystems are more difficult to conserve and why in all regions except Australia terrestrial conservation areas are more widely spread.

In response to concern about coral reef destruction, Conservation International also identified 11 marine hotspots including the Coral Triangle (Indonesia, Philippines and Papua New Guinea), which was subdivided into three hotspots: Indonesia, Sundaland and Wallacea. These hotspots contain 25% of the world's coral reefs and 34% of restricted range endemic species but cover only 0.02% of the oceans. Most of these marine hotspots are adjacent to terrestrial hotspots and experience land-based threats of pollution, over-fishing and tourism development. The marine hotspots encompass future planning and include some areas such as Chagos Island in the north Indian Ocean, which is currently a wilderness area with almost pristine coral reefs. Here the threats are more global, with bleaching from global warming and pursuit of specimens for ornamental fish collections in the Far East.

The value of ecosystems

Ecosystems are of enormous value to human wellbeing because of the range of services they offer. They provide the basic materials needed for subsistence such as food, freshwater, shelter, fuel etc. as well as contributing to human security (for example, mitigating the impact of disasters) and health (access to clean air and water). Figure 17 summarises the range of services ecosystems provide.

Researchers have attempted to quantify the direct and indirect values of ecosystems, but often **players** have very different views as to what is the greatest value of a particular ecosystem. This can lead to conflicts over their use.

Examiner tip

You are required to consider the value of a named global ecosystem such as coral reefs, rainforests or tropical grasslands, so this is a very important case study. Use Figure 17 as a framework.

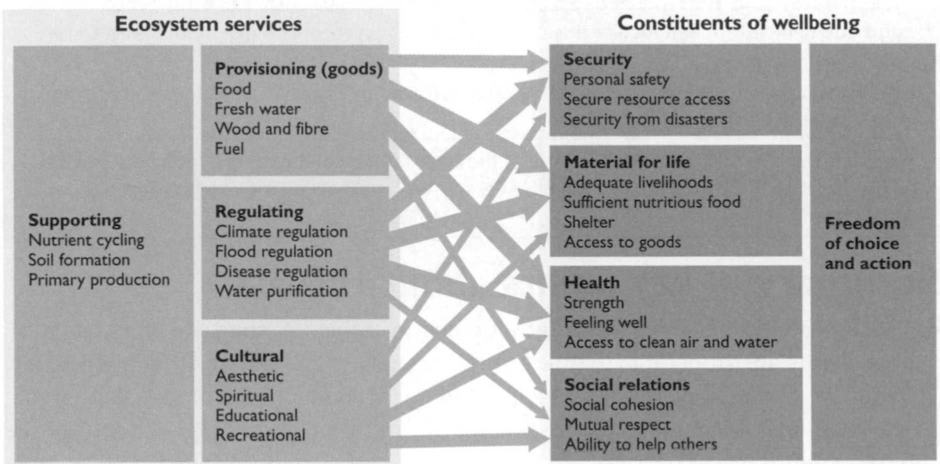

Figure 17 The value of ecosystem services

Threats to biodiversity

In order to assess threats to biodiversity it is necessary to audit the damage and destruction to them. Table 14 shows a number of different systems used to assess the threats, including websites for further research.

Examiner tip

Research the **www.cbd.int/gbo3** site for an extensive report on the state of biodiversity in 2010.

Table 14 Systems to assess threats to biodiversity

Organisation	Findings and methodology
WRI **www.wri.org**	**Ecosystem scorecard** which shows the condition of the world's major ecosystems and their ability to provide future goods and services. Records freshwater ecosystems as the most eco-stressed
WWF **www.panda.org**	**Living Planet Index** is developed by monitoring populations of representative **animal** species (319 marker species), initially in forests, freshwater (194 marker species) and marine ecosystems (217 marker species). Over time **Forest Index** fell by 12% in spite of an improvement in temperate forests (afforestation). **Freshwater** fell by 40% and marine by 30% but with signs of levelling off. **Grasslands** have recently been added, with tropical savannas/drylands showing spectacular decline (big game hunting, desertification). The decline in temperate grasslands is levelling off as most damage was pre-1970

Organisation	Findings and methodology
IUCN **www.iucnredlist.org**	Annual **Red List** of endangered species. Species are placed in one of ten categories ranging from extinct and endangered, to vulnerable, and to little concern. **Extinction hotspots** include tropical rainforests, tropical grasslands, polar environments and small island environments. Species with large body size are vulnerable to hunting for food. Species that have low rates of increase, poor dispersal and migration abilities and are easily predated by alien species or perceived as a nuisance by humans (e.g. rats and mice) are most vulnerable. **Freshwater ecosystems** have the highest percentage of threatened species, especially amphibians and reptiles
MEA United Nations Environment Programme (UNEP) **www.millennium assessment.org** or **www.unep.org**	This very large-scale survey across 13 major ecosystems reinforced existing findings: • **world drylands** (tropical grasslands) were under the greatest threat overall • **freshwater and marine ecosystems** were under greatest threat from hypoxia and eutrophication, and locally this could lead to ecosystem collapse. Coral reefs are particularly at risk • deforestation of rainforests was highly concentrated in Brazil and Indonesia. Together they cause 80% of the destruction
GBO — 3 (Global Biodiversity Outlook 3) **www.cbd.int/gbo3**	Assesses the current status and trends of biodiversity to celebrate the International Year of Biodiversity in 2010. The Year of Biodiversity was marked by a world conference in Aichi, Japan

Summary of key points about threats

- The assembled evidence from the surveys in Table 14 suggests that while all ecosystems are vulnerable to threats, it is freshwater ecosystems that are possibly under the greatest overall threats and it is the drylands (tropical grasslands) that may be under the greatest threat in the future. Climate change is seen as the biggest future threat overall.
- In terms of species, there are particular concerns about amphibians and reptiles (1 in 3 species in danger of extinction), mammals (1 in 4) and birds (1 in 8).
- **Iconic** species such as pandas, whales and polar bears are over-emphasised (because of their ability to attract conservation funding) at the expense of **keystone** species such as bees, which play an essential role in the food web and in the provision of services (pollination).
- A particular concern are **endemic** species which are noted for their uniqueness and rarity (e.g. in the Galapagos Islands).
- Threats can be classified according to scale from global to local. **Global scale threats** are widespread and severe in their impacts and can have a long-term effect on ecosystem functioning (nutrient cycling and energy flows, see Figure 18). **Local threats** operate at a particular location and their impacts are frequently short term. Table 15 summarises the range of threats. Note that **regional scale threats** can occur from the spread and development of a local threat, or from the specific impact of a global scale threat (e.g. rapid global warming in the Arctic).
- Some of the threats such as the impact of natural hazards (floods, hurricanes etc.) can be regarded as physical but the vast majority of threats are either partly or completely brought about by human actions. Figure 18 shows how the direct causes of **destruction** (loss in quantity) and **degradation** (loss in quality), also known as direct drivers of change, are often the result of underlying or **root causes**. For example, indigenous peoples hunt wild animals for food, and cut down trees for fuel wood to burn on their cooking stoves, but the root cause is their poverty, which means they cannot buy food and fuel from markets.

Examiner tip

When reviewing the state of biodiversity look at all the maps and statistics carefully, as they are all measuring different things. Remember that some of the data collection methods are relatively unreliable and rely on sampling, and using volunteers as well as scientists.

Knowledge check 24

Give examples of destruction and degradation of forest ecosystems to distinguish between the two terms.

Table 15 Global and local scale threats to biodiversity

Global threats	Local threats
• Accelerated soil erosion • Desertification • Deforestation • Global scale pollution, e.g. acid rain • Ozone depletion • **Global warming** • Disease	• Hazard impact • Wildfires • Habitat conversion and fragmentation • **Over-harvesting/fishing** • Recreational impacts • Opencasting from mineral working • Eutrophication and local pollution • **Invasive (alien) species**

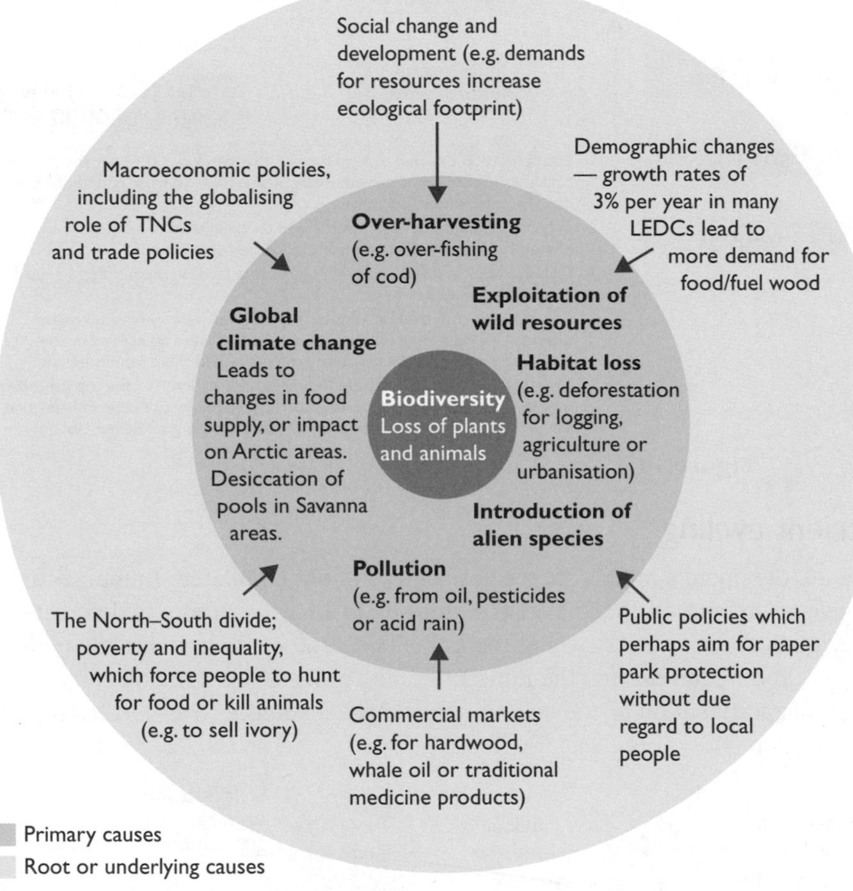

Figure 18 The drivers of ecosystems change

Examiner tip
You will need to assess the relative importance of threats. Criteria for assessment include the scale of threat, the damage potential, the immediacy of the threat, the length of the impact (short term or long term) and the ease of management. This assessment of threats should be supported by case studies by ecosystem, by location and by the type of threat. Use AS studies such as impacts of global warming in the Arctic.

The impact of threats on ecosystem processes

Energy flows

Figure 19 shows how energy flows through the trophic levels of an ecosystem. As energy is lost through respiration, at each stage of the process the amount of biomass at every trophic level decreases. Human actions can impact on any trophic level of the food web, which can in turn affect other trophic levels.

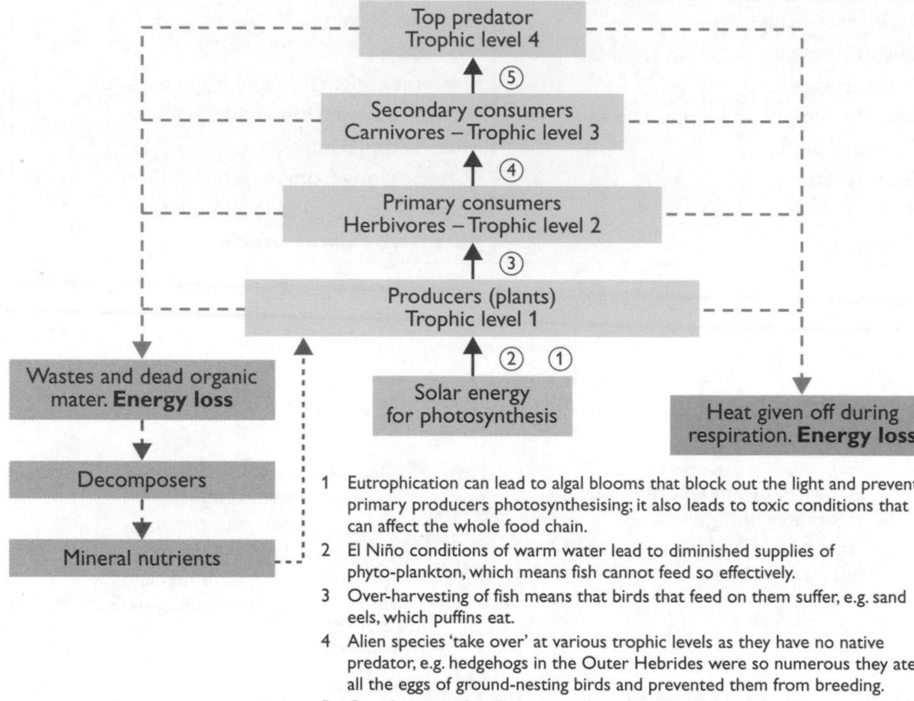

1 Eutrophication can lead to algal blooms that block out the light and prevent primary producers photosynthesising; it also leads to toxic conditions that can affect the whole food chain.
2 El Niño conditions of warm water lead to diminished supplies of phyto-plankton, which means fish cannot feed so effectively.
3 Over-harvesting of fish means that birds that feed on them suffer, e.g. sand eels, which puffins eat.
4 Alien species 'take over' at various trophic levels as they have no native predator, e.g. hedgehogs in the Outer Hebrides were so numerous they ate all the eggs of ground-nesting birds and prevented them from breeding.
5 Over-hunting of seals or otters has major implications for the top predators, whose source of food supply vanishes. Equally the loss of otters means that sea urchins proliferate and they in turn destroy the giant kelp forests.

Figure 19 Impact of threats on energy flows and food webs

Nutrient cycling

Figure 20, Gersmehl's model, shows how nutrients are circulated. Human activities can have an impact on the size of the store as in (1) where there is deforestation, or over-cultivation, or in (2) where there has been soil erosion. Indirectly pollution such as acid rain can destroy the nutrient cycling process (3). Humans can also add nutrients by fertilisation of the soil, in some cases leading to over-fertilisation and eutrophication (4).

Examiner tip

Choose two contrasting ecosystems and devise Gersmehl's models to show their differences in nutrient cycling.

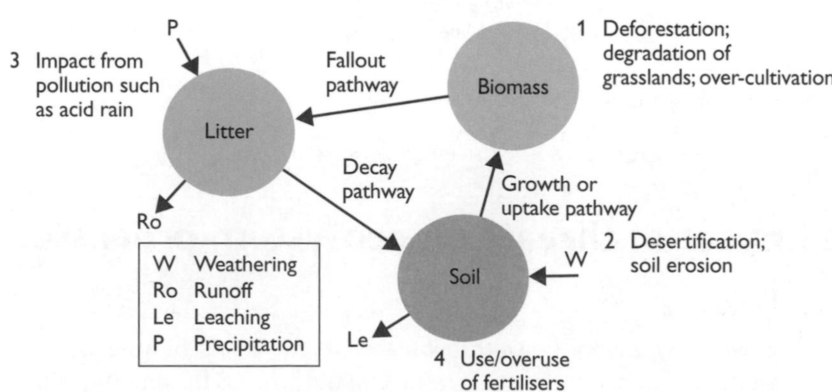

Figure 20 Impact of threats on nutrient cycling

Managing biodiversity

The concept of sustainable yield

As a result of exploitation, many ecosystems are under threat. The state of biodiversity is such that 2010 was declared the year of World Biodiversity in order to attempt to reverse the trend. **Sustainable yield** (see Figure 21) is a concept that is increasingly used as a key to successful ecosystem management, especially for marine ecosystems.

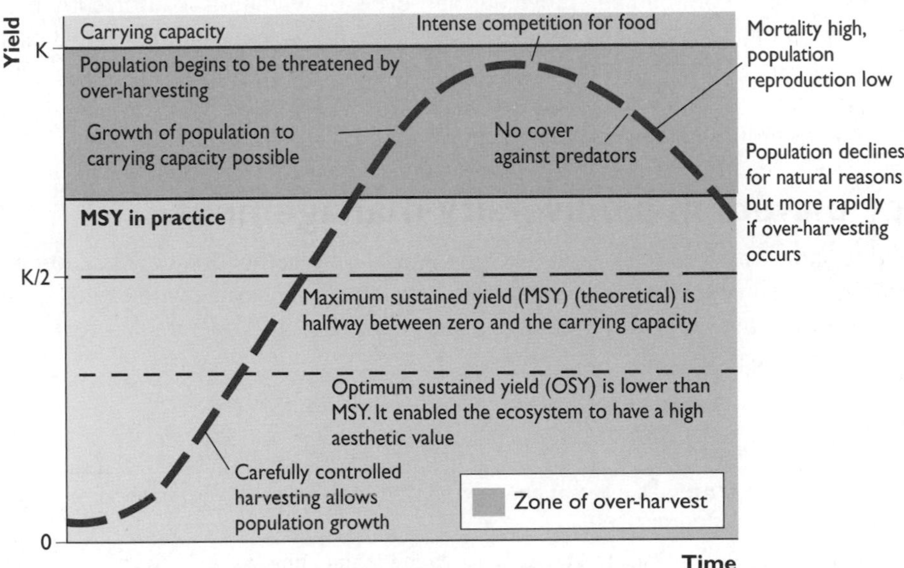

Figure 21 Sustainable yield

Sustainable yield is the 'safe' level of harvest that can be hunted/caught/utilised without detriment to the sustainable management of an ecosystem. Sustainable management allows the resource to be used but without compromising its worth for future generations.

- **Maximum sustainable yield** is the greatest harvest that could be taken indefinitely and leave the systems intact. Usually this is only exceeded by commercial activities.
- **Optimum sustainable yield** is usually used as a compromise because it is a lower level of yield and will not destroy the aesthetic or recreational value of the ecosystem but will allow multi-use for the maximum benefit of the whole community.

There are always problems in modelling the sustainable yield as there is frequently conflicting information on the numbers of species. The fishermen rarely agree with the researchers. There is also a problem in assessing populations in what is an environment without boundaries. Carrying capacity varies seasonally, and over time may be reduced by short-term climate change. As the carrying capacity of the habitat and the population numbers fluctuate, this makes modelling and management for conservation highly complex.

Knowledge check 25

Explain what the consequences are for ecosystems if the maximum sustainable yield is exceeded.

The Southern ocean fisheries north of Antarctica are an example where sustainable yield has been used very successfully. The only problem is illegal fishing from pirate boats, which in such a vast area is impossible to police.

However, sustainable yield becomes a controversial topic when applied to 'cute, cuddly' animals. It is used by managers to keep the trophic levels of an ecosystem 'in balance' — for example, by culling seals. Even more controversially it is used in **extractive reserves** — a concept pioneered by the CAMPFIRE scheme in Zimbabwe and now spreading to many game reserves in the tropical grasslands of Africa. Shooting and hunting for 'big game trophies' are permitted as a means of sustainable income for tribal communities provided that levels of wildlife are sufficiently high (i.e. above optimum sustainable yield) in order not to damage the parallel activities of wildlife tourism. While the idea of harvesting animals sustainably is becoming quite mainstream, many argue that 'licensed hunting' will never be a true conservation technique as profit will always get in the way.

Knowledge check 26

State the arguments for and against extractive reserves.

Key players in biodiversity management

Table 16 shows many of the key players who impact on whether and how biodiversity is managed. Some have negative impacts and are largely concerned with exploitation, whereas others are in the forefront of conservation. They also operate at a variety of scales from global to local.

Table 16 Key players in biodiversity management

Players	Roles
Individual farmers	Manage the environment for food consumption, now subsidised to deliver environmentally-friendly farming, adding to biodiversity
Individual campaigners	Campaign for justice to save the rainforest, e.g. Chico Mendez and Sting in the Amazon
Individuals consumers	Make decisions about their consumption, whether to use environmentally-friendly products or be ecotourists to conserve biodiversity
Artists, painters, writers	Provide images and literature to make people aware of issues, e.g. as Rachel Carson did on the degradation of the marine environment
Local communities	Develop sustainable management and work together to save local ecosystems using bottom-up schemes to resolve conflicts, e.g. Soufrière coral reef or the mangroves of Mankato Reserve in St Lucia
Campaigning NGOs	Greenpeace carried out specific campaigns, e.g. to Save the Whale. RSPB is the largest environmental organisation in Europe — campaigns to conserve environments for birds
Other NGOs' scientific research	Numerous NGOs support conservation and sustainable management schemes such as WWF in Udzungwa Forest, or Project Seahorse which carried out scientific research to manage marine environments
Local governments	In the UK they manage local reserves in cooperation with the Wildlife Trust and support Biodiversity Action Plans
National governments, e.g. in UK	There are various departments, such as DEFRA, which aim to provide a well-managed countryside and conserve wildlife. Natural England is responsible for designating and administering National Nature Reserves, SSSIs and AONBs
Transnational companies	Develop industries transnationally. All have environmental reports and their decisions concerning locations to exploit can have a huge impact on biodiversity in rainforest locations (logging/mining/biofuels). Collecting wild plants for pharmaceutical research can affect the gene pool. TNCs can sponsor biodiversity/conservation schemes and promote biodiversity

Players	Roles
International organisations	Global NGOs such as World Conservation Union and WRI (World Resources Institute) operate as think tanks and provide data and also administer world schemes such as Biosphere reserves and World Heritage sites
UNEP, UNESCO	Carry out research, exchange ideas, provide information on biodiversity. Were instrumental in the Millennium Ecosystems Assessment. Set up conferences for World Treaties such as CITES or administer agreements such as UNCLOS
Scientists and researchers	Work for a variety of organisations from NGOs to international organisations and agencies to provide research data for management of biodiversity, often based in universities

The results of the decisions these players make can lead to conflicts especially between exploitation of biodiversity for its goods, and conservation for its future use (for example, current and future use of the gene pool). Maximum rates of exploitation are occurring in countries experiencing rapid economic growth such as China.

Examiner tip

Devise two tables:
(a) to show which scales key players operate at and whether they have positive or negative impacts
(b) to summarise their relative importance

Conservation controversies

What to conserve

Assuming that globally, internationally and nationally only limited funds are available for conservation, decisions need to be made.

- A **hotspot** strategy prioritises the most threatened, high-quality areas, whereas an **eco region** strategy conserves representative ecosystems.
- A lot of fundraising favours high-profile iconic species (the cuddly toy syndrome) at the expense of vital keystone species and the future of the gene pool.

Knowledge check 27

Name the developers of the hotspot and eco region concepts.

Knowledge check 28

Name three NGOs involved with biodiversity conservation.

How to conserve

- Traditionally, total protection was favoured — this meant that any designated reserves were closed to all except a handful of scientists and local people could only obtain access illegally (and frequently did). It became apparent in the 1990s that this was not working effectively and so **sustainable development** became the way forward. Figure 22 summarises this spectrum of conservation. The result is a patchwork of all different types of reserve, with many different designations, but still only covering some 20% of possible areas of high value. There are huge variations in the numbers and size of protected areas in different countries.

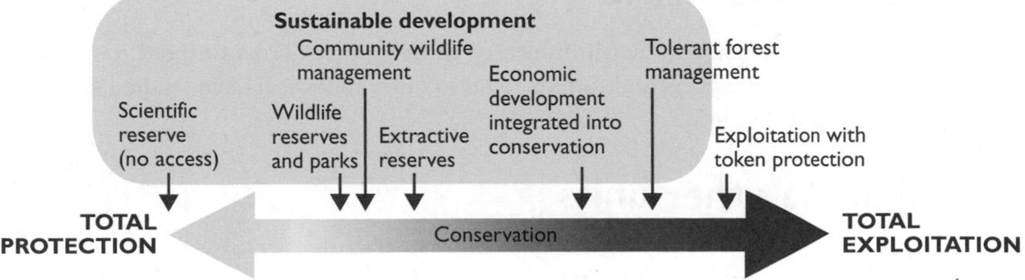

Figure 22 The conservation spectrum

- The actual nature of the eco reserves is very important. With the advent of global warming most designated reserves are much larger and allow space for

Knowledge check 29

Outline the advantages and disadvantages of several small or one single large reserve.

North–South migration (the debate is known as the SLOSS debate — Single Large Or Several Small).

- On eco reserves there is also a need for in-situ conservation to work alongside ex-situ conservation (e.g. seed banks, zoos, gene banks) in order to save endangered species from extinction. Historically, some zoos have experienced bad publicity with their conservation role underrated.

Mechanics

Inevitably there is potential conflict between top-down global strategies and some local initiatives and actions. Many indigenous peoples value biodiversity, and rely on it for their subsistence and therefore resent schemes imposed from afar. This has hopefully been lessened by detailed plans resulting from the 2010 World Year of Biodiversity, which integrate planning and policies at a variety of scales. There are also tensions between developed and developing countries as to who pays for the escalating costs of conservation. Costs tend to be lower in developing countries, but much of the money comes as aid and investment from developed countries.

Eco futures

Table 17 summarises four major scenarios for the future of biodiversity from three different organisations.

They vary in their degrees of sustainability and cooperation between nation states.

Examiner tip

Research the websites and rank the scenarios for each in terms of their likelihood of conserving the future of biodiversity.

Table 17 Biodiversity scenarios

Organisation	Scenario 1	Scenario 2	Scenario 3	Scenario 4
MEA **www.unep.org**	Global orchestration	Order from strength	Adapting mosaic	Techno garden
Living Planet Report **www.panda.org**	Business as usual	Slow shift scenario to sustainability	Rapid reduction scenario, radical solution	Shrink and shave
GLOBIO **www.unep.org**	Markets first	Policy first	Security first	Sustainability first

Synoptic links

The links below show how **biodiversity under threat** links to the three synoptic themes (players, actions and futures), and to other units you have studied as well as to wider global issues.

Links to other units

Unit 1

- **World at risk**: impact of climate change on Arctic and other ecosystems.

Unit 2

- **Crowded coasts**: impacts on SSSIs such as sand dunes.

Unit 3

- **Water conflicts**: diminishing water quantity and quality, impact of energy exploitation. Impacts of technology use, impacts of development on ecosystems.

Unit 4

- **Cold environments**: impact on fragile ecosystems in the Arctic.
- **Life on the margins**: impact of desertification on grassland ecosystems in drylands.
- **Consuming the rural landscape**: impact of tourism on biodiversity, especially in fragile and vulnerable areas such as Galapagos.

Links to wider global issues

- **Conservation vs exploitation** can be weighed up via biodiversity and the need for a sustainable approach can be considered.
- **Global warming** and its impact on biodiversity is a key concern especially for indigenous communities who depend on biodiversity for existence.
- The **development gap** can be illustrated with contrasting issues and attitudes to biodiversity.

Knowledge check 30

Explain what is meant by an SSSI.

Summary

- Biodiversity encompasses variability within and between species, and within ecosystems and their habitats.
- Biodiversity levels vary widely across the globe as a result of physical and human factors.
- Physical factors, especially climate, which controls the presence or absence of limiting factors, explain the big picture. Human factors have both direct and indirect influences on biodiversity levels at all scales.
- Human influences are seen as largely negative, but can also be positive through management and conservation.
- Biodiversity is of enormous value to the planet as ecosystems provide services that are vital to human wellbeing.
- All means of measurement of biodiversity, e.g. the IUCN Red List, suggest that in spite of considerable efforts, both the quantity and quality of biodiversity are under threat of destruction and degradation from a range of causes.
- These threats vary in scale, immediacy, length and severity of impact, and difficulty to manage.
- A number of key players are involved in the future of biodiversity at all scales from global to local. Some, such as TNCs, are largely exploitative, whereas others are in the forefront of conservation.
- There are many controversies about how best to conserve biodiversity, as there are limited funds available. For example:
  - Should you conserve hotspots (the areas of biodiversity under greatest threat) or representative eco regions?
  - Should we move away from total protection towards sustainable development or extractive reserves, which allow local people to participate and benefit?
- A number of future scenarios for the world's biodiversity have been developed. The most optimistic of these are based on sustainable development and a high degree of coordination between the key players.

Superpower geographies

Superpower geographies

What is a superpower?

Superpowers are countries with disproportionate power and influence. They are usually large countries in terms of population and physical extent. Physical size may provide a natural resource base that the superpower can draw on. True superpowers have global influence — the world is their backyard. Currently, there is only one superpower, the USA (Table 18).

Table 18 The USA versus the rest of the world in 2008

	The USA	Rest of the world	USA as % of world
Population	305 million	6.5 billion	4.6%
GDP	$13.8 trillion	$40.8 trillion	25%
Military spending	$711 billion	$759 billion	48%
Number of the world's 2000 largest TNCs	598	1,402	30%
Carbon dioxide emissions	6 billion tonnes	27 billion tonnes	22%

Examiner tip

Learn a definition of a superpower for the exam, and be prepared to justify your claims as to which countries are superpowers, emerging superpowers and regional powers.

One definition of a superpower is: 'a superpower must be able to conduct a global strategy,... to command vast economic potential and influence and present a universal ideology' (Professor Paul Dukes, University of Aberdeen). This definition emphasises:

- the global nature of superpower status
- the economic power of superpowers
- the political and cultural viewpoint or vision (ideology) that a superpower projects on the world

Superpowers have economic, cultural and military power resulting in global political influence. It is worth noting the other types of powers that are not as influential as superpowers.

- **Emerging superpowers** have growing influence, e.g. China. China has a growing economy and military power but less cultural influence than the USA. The EU has considerable power, but EU decisions and policies are often compromises because they must be the collective view of 27 member states.
- **Emerging powers** are further away from superpower status. They include a re-emerging Russia, India, perhaps Brazil and the Persian Gulf States. They have powerful cards such as energy resources, but not a full deck of global influence.
- **Regional powers** play an important economic and political role on their continent — examples include South Africa and Japan.

Knowledge check 31

Using data, outline the USA's claim to be the world's number 1 in terms of power.

Maintaining power

International influence has to be maintained. Superpowers do this using different mechanisms, some of which are hard (overt) and some of which are soft or more subtle (Figure 23).

Hard power ⟷ **Soft power**

Military presence and force
- Large air, naval and land forces
- Nuclear weapons
- Military bases in foreign countries giving geographical reach
- Military alliances such as NATO
- Diplomatic threats to use force if negotiation fails, and the use of force

Aid and trade
- Favouring certain trade partners by reducing import tariffs
- Trade blocs and alliances
- Providing allies with economic and technical assistance
- Using aid to influence policy or keep allies happy
- Using economic sanctions against countries

Culture and ideology
- Using the media to promote a particular image and message
- Exporting culture in the form of film and TV, or globally recognised brands
- Gradually persuading doubters that a particular action or view is in their interests

Figure 23 Mechanisms of power

Examiner tip
Consider which aspects of maintaining power (economy, military, culture, international decision making) are most important to different countries.

Superpowers use hard power mechanisms because these are the most obvious and threatening. The USA has an enormous military reach around the world giving it more military power than any other nation (Figure 24). Its military are present on every continent except Antarctica (but the USA has kept a permanently manned scientific base at the south pole since 1957 and in 2003 opened a new US$150 million base, reinforcing its superpower credentials). The NATO (North Atlantic Treaty Organization) military alliance provides the USA with allies in North America (Canada), Europe (UK, France, Italy and others) and the Middle East (Turkey). NATO was important during the Cold War period when the USSR was considered the USA's superpower enemy.

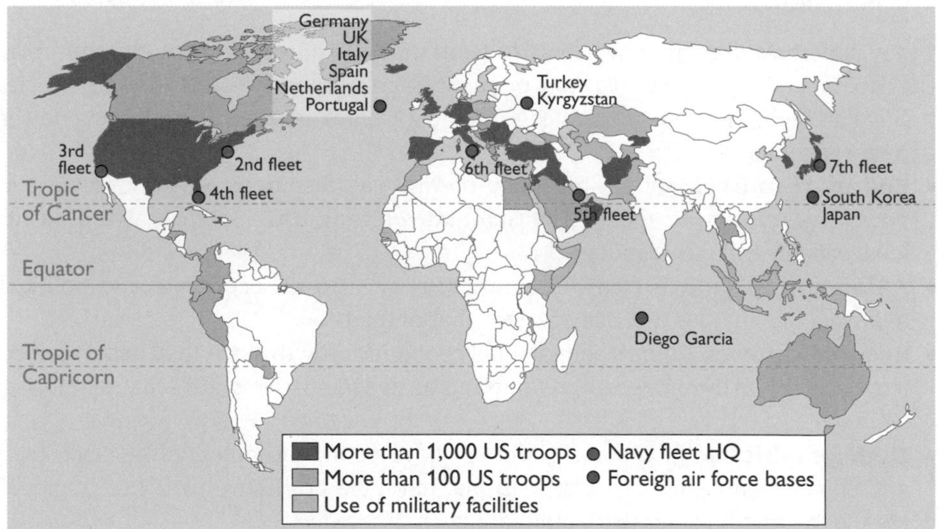

Figure 24 US military forces worldwide

Changing geographies

The geography of superpowers changes over time as old superpowers decline and new ones emerge. The number of superpowers in existence at any one time can also change (Table 19):

- A uni-polar world is one dominated by one superpower.

- A bi-polar world is one where two opposing superpowers exist.
- A multi-polar world is one with three or more superpowers.

Table 19 Timeline of changing superpowers

Timescale	Superpower(s)	Comment
1800–1918	British Empire	The UK is the dominant global power, at one point controlling around 25% of the land area of the Earth
1918–45	Transition period	Increasing power in the USA and Russia, the rise of Nazi Germany and the maintenance of the power of the British empire; arguably a multi-polar period
1945–90	USA and USSR	The Cold War period when the capitalist USA and the communist USSR squared up to each other for global domination
1990–2010	USA	Following the collapse of the USSR and eastern European communist states in the early 1990s, the USA has become the only true superpower (sometimes called a 'hyperpower')
2010–	USA, EU and China	Many observers think the future will be multi-polar, with many superpowers, possibly including India and Russia, jostling for power

After the Second World War, the British Empire declined quickly. By 1970 it had all but ceased to exist. This happened because:

- colonial countries demanded independence and political movements such as Gandhi's in India proved difficult to resist
- the UK could not afford to run a global empire, as the war had bankrupted the country
- there was a need to focus on postwar rebuilding and renewal in the UK, rather than in far-flung colonies

Knowledge check 32

What sort of geopolitical world existed between 1945 and 1990?

Today superpower geography seems to be in a period of transition as the uni-polar USA dominated world gives way to a more multi-polar one. The BRICs (Brazil, Russia, India and China) increasingly challenge the post-Cold War order. This growing power can be explained by:

- **Energy resources**: Russia has huge oil and gas resources, giving it economic power and 'energy weapons'. The Gulf States (Qatar, UAE, Saudi Arabia, Kuwait) have vast oil and gas resources.
- **Alliances**: EU growth from 6 countries in 1957 to an economic and political alliance of 27 in 2009. EU GDP exceeds that of the USA.
- **Economic power**: China's phenomenal economic growth since 1990 has propelled it to become the third largest economy and turned it into the world's manufacturing workshop.
- **Demographic weight**: some countries have economic potential because they are 'demographic superpowers' — China and India both have over 15% of global population and huge market potential.
- **Nuclear weapons** give countries power because they represent the ultimate threat. The USA, Russia, China, France and the UK all possess them. These countries are recognised as nuclear states under the Treaty on the Non-proliferation of Nuclear Weapons. Pakistan, India, Israel and North Korea are not signatories to the Treaty but have nuclear weapons.

Examiner tip

You do need to learn a timeline of superpower change and be able to refer to the changing geography of power in terms of uni-, bi- and multi-polar systems.

Theories

Several theories are used to explain the existence of rich, powerful countries and the weaker, poorer countries they dominate. The theories fall into two categories:

- Liberal economic development theories emphasise the creation of wealth and power and view capitalism as an essential tool for creating wealth.
- Marxist or Structuralist theories emphasise how some countries maintain their wealth and power at the expense of others; these theories see capitalism as promoting inequality.

Table 20 examines some competing theories.

Examiner tip
Don't forget to use theories in the exams. They can provide a useful structure for your answers. You will find many of these theories also relate to development (see Figure 29, p. 59).

Table 20 Theories explaining superpower status

	Theory, author and date	Explanation	Criticisms
Liberal	The take-off model (W. W. Rostow, 1960)	• Economic development is a linear, five-stage process • Countries 'take-off' and develop when pre-conditions are met, such as transport infrastructure • Industrialisation follows, creating jobs, trade and consumers	Many countries borrowed heavily and invested the money into projects to meet Rostow's pre-conditions, yet failed to develop and instead ended up in debt
	The Asian model (World Bank, 1993)	• Countries like China, South Korea and Taiwan have developed rapidly since 1970 • This is because they have opened up to free trade and foreign investment • The state has invested in education and skills development	The model fails to take full account of the support and aid provided to some Asian countries by the USA during the Cold War. Early in their development, many NICs had protectionist, not free-trade policies
Marxist	Dependency theory (A. G. Frank, 1967)	• The world is divided into North vs South • The developed world keeps the rest of the world in a state of underdevelopment, so it can exploit cheap resources • Aid, debt and trade patterns continually reinforce the dependency	Since the 1960s NICs and RICs have broken out of the North–South divide mould. The theory does not allow for developing countries to have a say in their own development
	World systems theory (I. Wallerstein, 1974)	• The world is divided into core, semi-periphery and periphery • Semi-periphery nations are broadly equivalent to the NICs that developed in the 1970s • Wallerstein recognised that some countries could develop and gain power, showing that wealth and power were fluid not static	World systems theory is more a description of the world than an explanation of it. It does not account for the rise of China and was written during the Cold War (bi-polar era)

India and China are both emerging powers, if not emerging superpowers. Importantly, neither country was allied to the Cold War superpowers and they have chosen to develop in their own way.

- Both countries largely avoided dependency by developing internally and shunning world trade until the early 1990s.
- Both invested heavily in home-grown technology, for instance space programmes, nuclear weapons and pharmaceuticals, without relying on other countries.
- Both opened up to foreign direct investment (FDI), and free trade focused in special economic zones, but only after a certain level of development was achieved.

Knowledge check 33
Does Frank's dependency theory or Wallerstein's world systems theory best fit the current world geopolitical situation?

Knowledge check 34

What factors hold India back from emerging as a world power?

India has developed some world-class software and IT companies of its own, such as Wipro and Infosys. Critics of India point to a lack of investment in basic infrastructure leading to poor transport and power shortages. In contrast, China has invested huge sums in infrastructure and is arguably better placed to attract FDI.

The role of superpowers

Control

Maintaining control over subservient colonies during the colonial era was relatively easy. The colonial model of direct control involved:

- using, or threatening to use, military force
- imposing government systems, usually run by administrators from the home country
- imposing the laws and sometimes language of the colonial power
- creating a different legal and social status between the colonisers and colonised

Examiner tip

There is a major difference between direct colonial control and indirect neo-colonial mechanisms of maintaining influence.

This model served European imperial powers such as the UK, France and the Netherlands well until the end of the Second World War. Colonies provided raw materials for the colonial power. Mines, farms, railways and ports were developed but little else. In the period 1945–80, many colonies experienced uprisings against colonial rule, such as the Mau-Mau rebellion in Kenya, 1952–60.

The expectation of former colonies was that they would quickly develop following independence. In fact this often failed to happen, especially in Africa. Left-wing thinkers explain this lack of development by arguing that direct colonial rule was replaced by indirect forms of control — or neo-colonialism (Table 21).

Table 21 Mechanisms of neo-colonialism

Developed nations	TNCs	International organisations
• Aid for corrupt dictators, such as President Mobuto of Zaire, in return for political support • Bilateral aid deals that benefited supplier companies in developed countries • Continuation of importing cheap raw materials and exporting expensive manufactured goods to the developing world • The brain-drain of skilled workers to the developed world	• Exploitation of developing resources, such as Nigerian oil, paying little in royalties and exporting profits • Protecting much needed technologies with patents and costly licensing agreements • Exploiting workers in low-skill factories by paying low wages	• Unsustainable lending leading to the debt crisis • Intervening in the economies of the developing world using structural adjustment policies to ensure debt repayments • Not doing enough to create a level trade playing field, which might increase trade with the developing world

Knowledge check 35

What is meant by 'neo-colonialism'?

Neo-colonialism falls into the Marxist/Structuralist camp of economic theory. Critics of the idea of neo-colonialism point to a number of factors missing from Table 21:

- Many developing countries have suffered long-term war and conflict which have prevented investment and development.
- Many NICs and RICs have developed following independence, especially in Asia but also in Latin America.
- Corruption has never been defeated in many developing countries and this ensures development finance and aid rarely reaches those who need it most.

International decision making

The most powerful countries are often members of overlapping international organisations, many of which have their roots in the 1940s. World leaders are often seen at meetings of organisations set up at, or immediately after, the Bretton Woods conference in 1944:

- the World Bank and International Monetary Fund (1944) — the twin global financial institutions; the USA has 17% of all votes on the IMF board
- the United Nations (1945) — the global organisation that promotes peace and cooperation; the five permanent members of the UN Security Council (UK, USA, France, Russia, China) have significant power
- the World Trade Organization (1947) — the global organisation that promotes free trade

Some economists argue that the WTO works mainly in the interests of the developed world.

Examiner tip
For the exam, you need examples of the influence of several different global organisations. These might include an economic organisation such as the IMF, a trade organisation such as the WTO and a military organsiation such as NATO.

In addition to these global organisations, developed countries have created other inter-governmental organisations (IGOs) for the purposes of maintaining power. Trade blocs, for instance NAFTA and the EU, promote free trade between member countries but impose tariffs and quotas on those countries outside the bloc. The EU is an example of a supra-national body. The 27 EU countries have a number of joint policies in areas such as:

- economics and free trade (the single market)
- foreign and defence policy
- social and environmental policy

By creating a powerful political and economic grouping, the EU has increased the power of European countries. The EU 'beats' the USA on many measures, as shown in Table 22. The EU is less clearly a superpower because it is still troubled by internal arguments and differences: France has objected to Turkey's proposed membership and the UK has not joined the euro.

Table 22 The USA and EU compared

USA	Measure (2010 data)	EU
308 million	Population	501 million
14.6 trillion	Total annual GDP (US$)	15.2 trillion
47,100	GDP per capita (US$)	30,400
687 billion	Annual military spending (US$)	400 billion
162	Number of world's 500 largest TNCs	170

Examiner tip
The EU is an interesting example. Is it a superpower or not? Be prepared to compare its strengths and weaknesses with those of the USA.

Americans are on average richer than Europeans, and the USA has a more powerful military machine. In other respects the two are similar. The USA, UK, France, Germany and Italy are members of the G8 (Group of 8) along with Canada and Japan. Russia has attended G8 summits since 1997, reflecting Russia's oil and gas wealth. G8 summit meetings are the ultimate international organisation because the leaders at each meeting represent:

- 65% of global wealth

Examiner tip
Learn some key facts and statistics for the exam. They make your work precise and detailed.

- 95% of global nuclear weapons
- 75% of global military spending

Interestingly, G8 nations make up only 15% of world population.

Trade

Wealth and power often seem one and the same, but where does wealth come from? Trade — the exchange of goods and services — generates wealth. Table 23 shows that world trade in goods (imports and exports) is dominated by trade within and between certain regions. Five trade flows accounted for 63.7% of all world trade in goods in 2007. Africa accounted for just over 2% of all world trade in goods in 2007, versus over 40% for Europe.

Table 23 Inter- and intra-regional trade in goods, 2007

Key: Over 5% · 1%–4.9% · Under 1% · No data = 0.1% or less

		Destination						
		North America	South and Central America	Europe	Russia/Central Asia	Africa	Middle East	Asia
Origin	North America	7.7	0.9	2.4		0.2	0.4	2.7
	South and Central America	1.1	0.9	0.7				0.5
	Europe	3.7	0.6	31.0	1.2	1.0	1.1	3.1
	Russia/Central Asia	0.2		2.1	0.7			0.4
	Africa	0.7		1.3		0.3		0.6
	Middle East	0.6		0.9		0.2	0.6	2.9
	Asia	6.0	0.6	5.1	0.4	0.6	0.9	13.9

Knowledge check 36
Which continent is the least involved in world trade?

The superpowers and emerging powers have a significant advantage when it comes to world trade:

- most TNCs originate in rich, developed countries
- most major shipping companies and airlines originate in the USA and Europe
- many developed countries allow free trade between each other

Examiner tip
Trade, and the money it generates, is crucial to superpower status. For this reason it is important to have some key facts and figures to hand that relate to China, the USA and the EU.

Less powerful countries export low-value commodities such as coffee, copper and cotton to developed countries, but import costly manufactured goods. Long term, the value of commodities has fallen relative to the cost of goods and services, which has worsened terms of trade. Some emerging powers have broken out of the commodity export model. China is now a major exporter of high-value manufactured goods and India has moved into IT and software exports. These are both wealth creators, helping to explain rising wealth and power. Russia and the Gulf States use oil and gas as their trade weapons. State-owned companies, e.g. Saudi Aramco and Gazprom, control export volumes and profits.

Superpower culture

Arguably the biggest superpower export is culture. People from the same culture share common values and beliefs, which may include:

- religious beliefs, attitudes, moral and ethical values
- shared language, dress and art (music, dance, literature) and symbols
- various 'norms' such as ways to behave and commonly accepted laws

Many geographers recognise the emergence of a global culture spreading around the world. This is a result of cultural globalisation. As global culture is seen as being dominated by Europe and North America, the term 'Westernisation' is used (plus Americanisation, McDonaldisation, Disneyfication or Cocacolonisation). Table 24 shows the important characteristics of the Westernised global culture that the USA in particular is accused of exporting.

Examiner tip

Cultural influence is a good example of 'soft' power, which can be contrasted with the 'hard' power provided by military force.

Table 24 Western values?

Value	Definition
Democracy	The belief that a developed society is one where everyone has the right to vote
Individualism	The belief that individuals should have the right to pursue their own actions and dreams
Consumerism	The belief that wealth, and the ability to buy goods and services, leads to happiness
Technology	The belief that problems can be solved by using technology, especially high-end technology
Economic freedom	The belief that markets should be free, and people should be at liberty to make money how they choose

Critics of cultural globalisation argue that as Western culture spreads it threatens traditional cultures and eventually dilutes them. It is difficult to see how cultural globalisation might be prevented as many factors promote it (Figure 25).

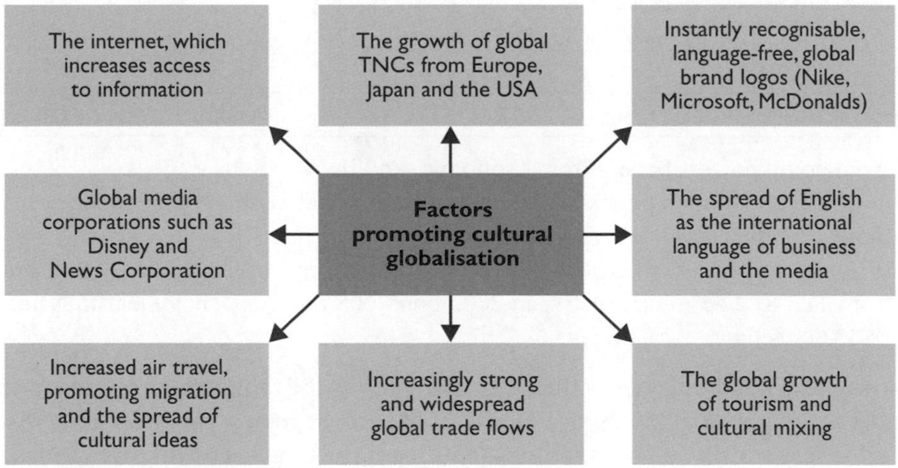

Figure 25 Factors promoting cultural globalisation

Fears that a global culture will destroy local cultures has led to a backlash against Western, especially American, cultural values. This backlash is complex as its cultural aspect is clouded by people's attitudes towards the Iraq War, large American corporations and even the environmental agenda.

- Direct action protests against globalisation occur regularly — for instance the 1999 WTO meeting in Seattle, the 2001 G8 Summit in Genoa and the 2005 Summit at Gleneagles.
- The 2005 UN Convention on the Protection and Promotion of the Diversity of Cultural Expressions allows countries to use measures to protect their culture by, for instance, limiting imports of foreign media.

Knowledge check 37

Why do some groups of people dislike the trend towards a Westernised global culture?

Cultural globalisation is not accepted by everyone. Globalisation could be said to promote cultural mixing, multiculturalism and better cultural understanding. The internet allows people to keep in touch with each other and share and maintain cultural values.

Superpower futures

The BRICs

Brazil, Russia, India and China are all emerging powers and potential superpowers. The Persian Gulf States have aspirations too. In the future, the USA and EU may be less powerful, but they are unlikely to give up their top slots easily. As can be seen in Table 25, China's economic growth outpaced the 'old' powers between 2000 and 2007. The BRICs are rapidly adopting communication technology, and improving life expectancy through better healthcare.

Knowledge check 38

What evidence is there in Table 25 of the rapid adoption of new technology by emerging powers?

Table 25 Change in the old and new powers, 2000–08

Country or area	Total economic growth 2000–07	Internet use growth 2000–08	Additional years life expectancy added 2000–05	% of world population 2007	% of world GDP 2007
EU	90%	210%	+0.9	14%	67%
USA	41%	131%	+0.8		
Japan	−6%	100%	+0.7		
China	185%	1024%	+1.0	42%	11%

The emerging powers benefit from economic growth:

- in China, 200 million people moved out of poverty between 1990 and 2005
- India's middle class has swelled to over 300 million
- in Brazil the number of households with an income of US$5,900–22,000 grew from 14.3 to 22.3 million between 2000 and 2005 and households earning under US$3,000 fell by 1.3 million

Examiner tip

It is very important to recognise that the BRICs are four very different countries, with contrasting strengths and weaknesses. In the exam you must discuss the separate countries rather than give a generalised discussion of the 'BRICs'.

The swelling middle classes of the BRICs are entering the ranks of global consumers. This has the potential to improve quality of life and happiness, but also to increase heart disease, obesity and stress-related illness. In China, growing wealth could increase pressure for political change, freedom and democracy.

A world of BRIC consumers is likely to be one of strained resources and environmental concerns. Projections of BRIC per capita GDP by 2050 are startling (Table 26) and suggest that by that time most will be consuming resources at similar levels to the UK and USA today. This would have major implications for water resources, land, air quality and ecological footprints.

Table 26 The BRICs in 2050?

	US$ GDP per capita, 2007	US$ GDP per capita, 2050	Ecological footprint, per person (global hectares), 2008	Additional population, 2007–50
UK	46,000	80,000	5.3	5 million
USA	46,000	90,000	9.4	120 million
Russia	9,000	78,000	3.7	–30 million
Brazil	7,000	50,000	2.4	130 million
China	2,500	50,000	2.1	150 million
India	1,000	21,000	0.9	500 million

Source: Goldman Sachs, 2007

Knowledge check 39

What evidence is there in Table 26 that the BRICs are a varied group of countries?

Out with the old and in with the new?

The BRICs may challenge the superpower status of the USA and the relevance of the EU. There are plans to change the UN Security Council — India and Brazil are serious candidates to get a permanent seat at the top table. Other threats to the existing powers include:

- **Resources**: as supplies of fossil fuels and rare metals like platinum begin to tighten due to increasing demand and a shrinking resource base, countries may find themselves in a bidding war to claim resources.
- **Military dominance**: three of the BRICs are now nuclear powers (Russia, China and India) and this could increasingly shift the balance of military might towards Asia.
- **Space**: once the preserve of the USSR, USA and European Space Agency, both China and India have active, well-funded manned space programmes to explore the final frontier.
- **Outsourcing**: India has benefited from software and IT outsourcing, and China from the global shift in manufacturing. If this continues and expands, jobs and prosperity in the West could be eroded.
- **Ageing**: Japan and much of the EU have increasingly ageing populations and face a pension funding crisis. This is also a problem for China (a result of its 'one-child' policy) and Russia. India and Brazil are much more youthful, and potentially innovative. The emerging powers have government-owned investment funds called sovereign wealth funds (Table 27). Much of the money comes from oil wealth.

Examiner tip

Don't forget about the demographic aspects of superpower status. The EU, Japan, Russia and China all have potential issues with ageing populations, whereas the USA, India and Brazil are much more youthful.

Table 27 Sovereign wealth funds, 2008

Country	Fund value, US$ billion (approximate)
Gulf States (combined UAE, Qatar, Kuwait, Saudi Arabia)	1,300
China	200
Russia	34

These funds are used to buy assets around the world, which then are effectively owned by a foreign government. This has become a cause for concern because foreign governments could control strategic assets in another country. The Abu Dhabi SWF owns 10% of the US bank Morgan Stanley; China's SWF owns a US$100 million stake in Visa credit cards and 5% of the US bank Citigroup; the Kuwait SWF owns 7.6% of Daimler.

Knowledge check 40

What is a sovereign wealth fund (SWF) and why do they trouble some existing powers?

Examiner tip
China's role in Africa is a developing story which you need to keep up to date with. The BBC news website is an ideal place to start your research.

China has recently found another way to extend its global influence by investing in oil and mineral extraction in Africa (Table 28). This is in the form of FDI or development assistance — usually infrastructure building. This has been dubbed a new form of colonialism. In fact, China seems to want guaranteed resource supplies rather than political and economic influence. In Angola, China has provided soft loans of up to US$4 billion in exchange for 10,000 barrels of oil per day. In Sudan by 2007, Chinese investment totalled at least US$1.3 billion — again in exchange for oil.

Table 28 China in Africa

New opportunities and benefits?	Continued dependency and problems?
• Chinese FDI has boosted Africa's economic growth • Africa has a new market for its raw materials • China may only be interested in resources, not controlling African governments • Chinese investment in mines and factories often includes new roads and other infrastructure • China is providing development assistance in exchange for goods, so countries avoid debt	• Chinese factory-produced goods undercut the price of local goods • Many employees are actually Chinese immigrants, so local Africans do not benefit from jobs • There have been some clashes between Chinese and Africans in Zambia and Equatorial Guinea • China has been accused of meddling in politics, for instance the Darfur conflict in Sudan, to protect access to its oil supplies

Rising tensions?

Many observers think that a new world order is emerging and that this is likely to be a multi-polar one (Figure 26). This could lead to a future of increased tensions.

- With no dominant superpower, and greater equality of power, sabre-rattling and argument may be more common.
- Some emerging powers have large energy resources that other powers will want.
- With no dominant power, trade and political agreements may become more bilateral (between two countries or regions) and less global.
- Bilateral agreements could increase tensions as some countries feel sidelined or excluded.
- Emerging powers may interfere in traditional superpower spheres of influence such as the Middle East.

Knowledge check 41
Explain what is meant by a multi-polar world in the future.

There are already many flashpoint locations where simmering problems could drag emerging powers and superpowers into conflict (Figure 26).

There is also the possibility of a 'clash of cultures' in a multi-polar world. While European and American cultures are similar, this is not true of Asian and Middle Eastern cultures. To some extent the clash of America and Islam has already happened:

- the 2001 9/11 bombings
- the 2003 invasion of Iraq
- ongoing conflict with the Taliban in Afghanistan
- the ongoing 'war on terror' against Al-Qaeda

Examiner tip
Remember that superpower futures are uncertain. You will gain marks in the exam by recognising this uncertainty and justifying your own view.

These conflicts and events have increased tensions between the Islamic and American worlds. In future, parts of the Islamic world that are oil rich may gain more wealth and power, increasing their confidence to treat the USA as an equal.

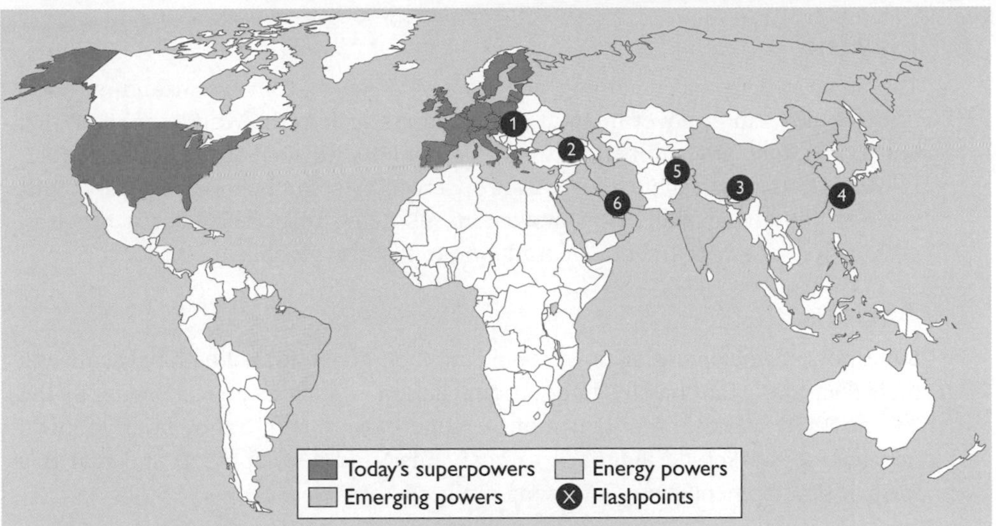

1 Ukrainian moves to join the EU or NATO may prompt Russia to disrupt gas flows to Europe
2 Georgia may press for NATO membership, increasing tensions in the Caucasus
3 China and India have an unresolved border dispute in South Tibet
4 The future of Taiwan, North Korea's nuclear weapons, and the dispute over the Kuril Islands between Japan and Russia are all potential conflict sources
5 The unresolved Kashmir dispute could lead to conflict between India and Pakistan
6 Iraq, a nuclear armed Iran and terrorism could all ignite tensions in the Persian Gulf

Figure 26 The 2030 multi-polar world?

The USA and Europe do not see eye-to-eye at all times. There are major differences in terms of:

- politics — Europeans view 'socialism' as a policy choice rather than an expletive
- social policy — in the USA individuals provide for themselves more, and the welfare state is smaller
- interventionism — the USA has been more ready than Europe to take military action to protect its perceived interests
- trade — free trade and free-trade agreements tend to be viewed more sceptically in the USA

These are all possible sources of increased tension. There is the thorny issue of Asia and how to tackle the crouching Bengal tiger and soaring Chinese dragon. So far the EU and USA have invested in both countries, increased trade links and maintained good relations. But for how long? Lastly, there is the Russian bear and its huge oil and gas reserves, vast central Asian sphere of influence, and borders with numerous political hotspots. In 2005–09 Russia has already shown how it can destabilise Europe by cutting off gas supplies — will this happen more in the future?

Synoptic links

The links below show how **superpower geographies** links to the three synoptic themes (players, actions and futures), and to other units you have studied as well as to wider global issues.

Players

The superpowers and emerging powers play a key role in world affairs, often through global inter-governmental organisations such as the World Bank, IMF, UN and WTO. Some are members of supra-national bodies such as the EU and OPEC. Powerful countries such as the USA are viewed by some political organisations, such as the anti-globalisation movement, as being responsible for a range of problems from environmental degradation to cultural globalisation.

Actions

Superpowers and emerging powers play out their geopolitics at a global scale and their decisions often involve international action — such as peacekeeping by the UN, or NATO warfare. A criticism of the superpowers is that they often follow a neo-liberal, market-led approach to economics and development, and that this form of development is not environmentally sustainable.

Futures

The future of the superpowers is not yet clear. A business as usual future would involve the USA maintaining its uni-polar dominance. Many economists and politicians think that a multi-polar future is more likely, with the BRICs emerging as powerful geopolitical and economic forces. Given the influence of superpowers, if one decided to move towards a more sustainable consumption model, the rest of the world might follow.

Links to other units

Unit 1

- **Going global**: globalisation plays a key role in the power that superpowers yield, and as a wealth creator.

Unit 3

- **Water conflicts**: the emerging superpowers of India, China and the Gulf States increasingly place demands on water resources, many of which are finite.
- **Energy security**: oil and gas is a key source of power for Russia and the Gulf States, and in future may be an increasing source of geopolitical tension.

Links to wider global issues

- The process of **globalisation** and the increasing connectivity of many parts of the world are driven by players in superpowers and emerging powers. These include TNCs, governments and consumers. The costs and benefits of globalisation — environmental, economic and social — can be seen clearly in China and India.
- Examining the contributions made to **global warming** by different countries around the world reveals that in 2007 the USA emitted 22.2% of all greenhouse gases, China 18.4%, the EU 14.7%, Russia 5.5%, India 4.9% and Brazil 1.2% — a combined total of 66.7%. The global warming problem originates in these countries, and any solutions will have to involve them.

Summary

- Strategically, superpowers are global players based on political power derived from economic, military and cultural strengths.
- Superpower status is maintained by overt (hard) mechanisms such as military might, and by more subtle 'soft' powers such as cultural influences (reinforced by global media and TNCs).
- Superpower geographies change as old superpowers decline and new ones emerge:
  – A bi-polar world existed between 1945 and 1990 with the Cold War between the capitalist USA and communist USSR.
  – A uni-polar world currently exists, with the USA as the only superpower.
  – A multi-polar world will almost certainly exist in the future with the emergence of the EU, China and possibly India, and the re-emergence of Russia.
- A range of liberal and structuralist economic development theories can be used to explain the development and existence of superpowers.
- Superpowers reinforce and promote their roles by participating in trade and international decision making, investing sovereign wealth funds (sometimes seen as neo-colonialism), developing a network of military bases, and facilitating the global spread of their culture.
- In the future, the likely occurrence of a multi-polar world, with emerging powers acquiring greater status, will lead to a more complex situation. China and USA could develop rival spheres of influence, leading to increased tensions.

Bridging the development gap

The causes of the development gap

Defining development

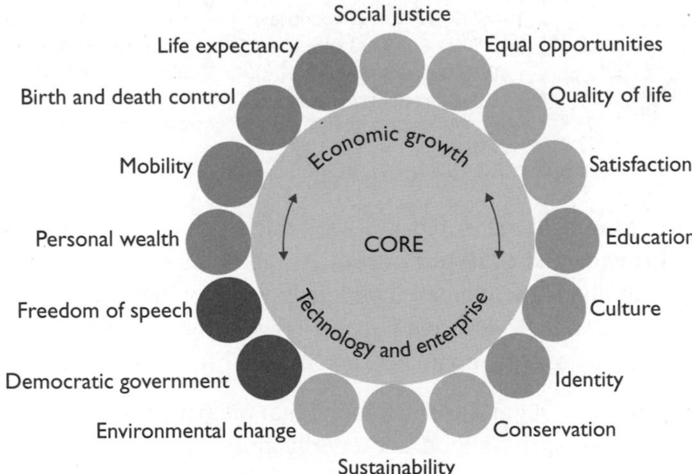

Figure 27 The development cable

Development implies change, growth or improvement over time. Initially development referred to a growth in wealth, i.e. **economic development**, but in the 1980s development began to focus on **human development** (standards of health, education etc.) and the improvement in people's quality of life. Figure 27 shows how the twenty-first century's rethinking of development has led to a holistic, all-round approach, which looks at economic, human, environmental, socio-cultural and political matters.

Knowledge check 42

Explain the terms Brandt line and North–South divide.

The **development gap** can be defined as the difference in income and quality of life between the richest and poorest countries of the world. The Brandt report (1980) identified this disparity which became known as the **North–South divide**.

Measuring development

Economic development is measured either by GDP per capita or GNP per capita. In order to make comparisons easier the raw results are converted into PPP (purchasing power parity) in US$.

Examiner tip

Redraw the Brandt line to reflect the current situation of wealth in the world. Summarise the changes you would make and why.

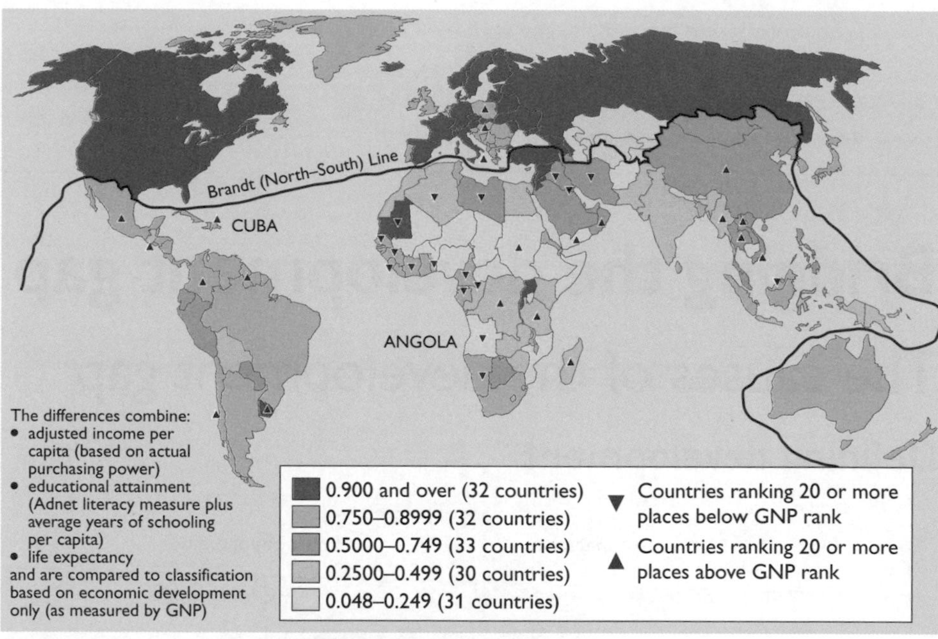

Figure 28 HDI and GNP per capita compared

Human development is measured by HDI, which is based on three factors: **life expectancy**, **literacy** and **GDP per capita**. As Figure 28 shows, although there is a clear North–South divide there is not a perfect correlation between wealth and HDI. The data in Figure 28 are 15 years old but, even though both GNP per capita and HDI have improved, there are still anomalies such as Cuba (+) and Angola (–).

Examiner tip

Use United Nations statistics to find out contrasts between selected countries, and variations in development indicators, so that you can support your assertions in exam answers.

Other indexes include PQLI and various individual measures to reflect the widening scope of development such as food and energy consumption, infant mortality and more recently GDI (gender development index) which measures the role females play in development. The technology attainment index (TAI) has been introduced recently to reflect the digital divide between the 'switched on' and 'switched off' world.

Development shows an unequal world, sometimes called the 80:20 world — the rich world enjoys 80% of the world's wealth but contains only 20% of the world's people.

What is happening to the development gap?

Evidence using a wide range of indicators to create **development profiles** of countries would suggest that the development gap:

- is **widening** at the extremities, with rich MEDCs continuing to get richer, whereas the world's poorest countries (known as LDCs, or low income countries) are at best remaining static or at worst getting poorer as a result of climate change, famines, wars, natural disasters etc. These LDCs are concentrated in the African continent.
- is **narrowing** elsewhere. Many developing countries are becoming RICs, or expanding tourism and commercial agriculture. Many countries, especially in the Far East, are joining the ranks of the NICs, including India and China (the two emergent superpowers). Singapore, an original first generation NIC, is now so wealthy that it is classified as an MEDC. Many of these countries are now members of the G20 group of countries (a world economic forum). Another group of countries, members of OPEC (including Gulf States), are also becoming very rich as a result of their oil wealth. The Former Communist Countries (FCCs) are an anomaly as many have stagnated or declined — especially the non oil-rich.

Examiner tip

Revise all your acronyms for groups of countries from AS Unit 1: Going global. Use the World Bank Reports to note the latest groupings of countries. Draw a spectrum of development as shown below and then summarise what changes in development have taken place.

MEDC HIC OPEC NIC FCC MIC RIC LIC LDC

High income Middle income Low income

Theories of development

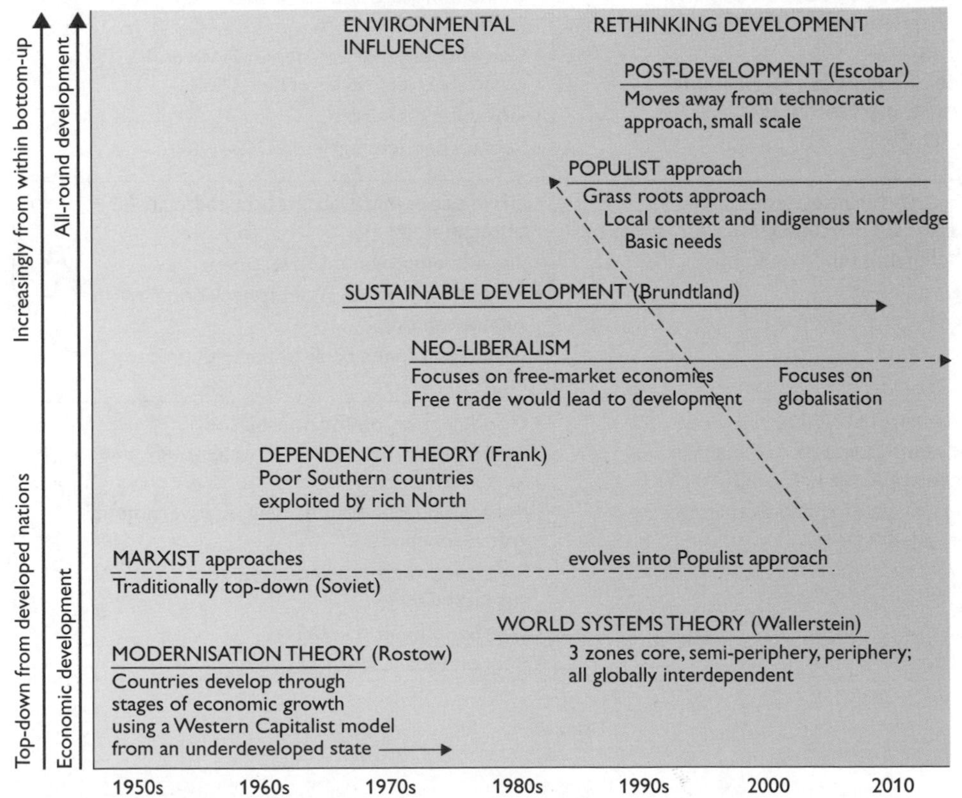

Figure 29 A chronology of development theory

Knowledge check 43

Describe and suggest reasons for the trends in development theory shown in Figure 29.

Figure 29 summarises the theories used to explain the development gap. It shows not only how the way we define development has changed in the last 70 years, but also how the theories that developed to explain the 'gap' have evolved over time.

Theories were initially developed by the rich North to explain why the South was impoverished, but now theories are increasingly coming from the developing countries themselves.

Table 29 The major organisations that have an impact on global development

Organisation	Role	Impacts	+/–
United Nations (development agency)	• Monitoring and managing investment via development agency, FAO, UNICEF etc. • Developed the Millennium Development Goals (MDGs)	Large-scale projects get the job done and problem solved, though sometimes slow to act as decision making tied to UN security council	+ –
International Monetary Fund (IMF) (UN agency)	• Manages financial transactions if countries get into debt • Renegotiates debt with strict conditions (stabilisation programme) • Responsible for HIPC initiative	• Stops a country going bankrupt and disruption of world systems • Stabilisation and structural adjustment programmes can hinder development	+ –
World Bank (UN agency)	• Borrows around US$20–30 billion to lend money to finance around 4,000 development projects in 130 countries (traditionally top-down)	• Provides investment for economic and social projects to improve life in many countries • Lends at market rate for top-down projects	+ –
World Trade Organization (WTO) (trade blocs)	• Promotes free trade and economic cooperation between countries, tariffs/quotas etc. • Groups of countries that promote free trade, e.g. NAFTA, EU, ASEAN, MERCOSUR	• Can promote trade (though not always fairly) • Can encourage trade dependency and create barriers to fairer/fair trade • Good for members • Sets up barriers against LDCs	+ – + –
TNCs	• Provide FDI for investment in many countries (see Fortune Global 500 index), e.g. Shell, Unilever, Nestlé, LG	• Drive economic globalisation and transfer of technology • Provide employment/investment • Possible exploitation of cheap labour by subcontracting • Leakage of funds back to parent company HQ	+ + – –
Global NGOs (BINGOs) Non-governmental organisations (NGOs) specialising in development	• Big international NGOs carry out government contracts and manage aid projects, e.g. Oxfam, CAFOD, WWF • Smaller specialist charities that carry out bottom-up schemes	• Good reputation for non-biased development assistance — emergency and long term • Reliant on donations as well as government grants for funds • Specialist, targeted and often very successful • Can be difficult to replicate • Rely on local capacity building	+ – + – +

Organisation	Role	Impacts	+/–
National governments, e.g. DFID UK	• Can influence development priorities in their country and provide overseas aid with bilateral agreements	• Can supply large amounts of much needed development aid, e.g. in big transport projects • Can be tied and linked to the ex-colonial government's priorities, i.e. politically motivated, also seeking economic benefits	+ –
University development departments	• Carry out research, development and monitoring projects for a variety of contracts	• Skilled advice at forefront of research • Can have biased views depending on funding source	+ –

As Table 29 shows, there are positive and negative impacts for all the actions of key players. Some, such as World Bank, IMF and WTO, attract huge opposition, especially from left-wing press, who claim that they actually hinder and impede development and may have contributed to the widening of the development gap. Other organisations, such as NGOs, are seen as saviours of the poor and are regarded favourably because of their ability to deliver bottom-up, sustainable projects that can be very effective locally.

For some other organisations, such as TNCs (link to AS Going global), there are both positives and negatives depending on the TNC itself and the location and nature of its activities in developing countries. Many TNCs are now working hard to change their image as exploiters of poor countries.

Examiner tip
Research the HIPC initiative and explain how it can be beneficial to LDCs.

Knowledge check 44
How many MDGs are there and when must they be completed by?

Examiner tip
Always check the bias of the books, articles and websites you use because development is a very polarised topic. You need to research a range of factual case studies on the players listed in Table 29.

The consequences of the development gap

The impact of the development gap has different consequences for:

- **different groups of countries** — the very poor countries are most disadvantaged and are experiencing the largely negative consequences of a widening development gap with worsening or at best static economies. Some developing countries are benefiting from economic growth and are significantly reducing poverty levels. However, economic development comes at a price as there are negative environmental and social impacts that are a challenge to manage.
- **different groups within countries** — as countries develop, rapid urbanisation occurs leading to the growth of megacities. In the early stages of the urbanisation of emergent megacities the traditional poverty of rural areas is transferred to urban slum areas as people move from the country to become the urban poor.

Within the population of every country there are disadvantaged groups, such as women, disabled, ethnic and religious minorities. They are bypassed by the positive impacts of development. As they experience a growing personal development gap this can lead to social unrest and the development of new political movements that try to fight the negatives of disparity. The 2011 Middle Eastern unrest has been largely fuelled by younger people, both men and women, many of whom lack opportunities for well-paid employment as well as freedom of expression.

Examiner tip
Research two contrasting LDCs in the continent of Africa and list the factors that have led to their poverty.

The consequences of the development gap can be classified as economic, social, environmental and political, and are largely negative (see Table 30).

Consequences for very disadvantaged poor countries, variously known as least developed countries (LDCs), low income countries under stress (LICUS) or heavily indebted poor countries (HIPCs), are summarised in Table 30. There are currently around 50 countries in these categories, largely concentrated in Africa.

Table 30 Consequences of the development gap for LDCs

Factor	Consequences
Economic	• Lack of money for development — lack of foreign investment • Lack of money to pay for food, investment in rural development • Lack of money to build infrastructure — limited transport, health, education • Poor employment prospects — largely dependent on a few primary exports • Isolated and switched off from connections and globalisation • Low personal incomes — many inhabitants living below the poverty line
Social	• Little formal investment in healthcare and education, low levels of literacy and skills, high levels of infant mortality, low life expectancy • Inability to combat pandemics such as HIV/AIDS • Basic infrastructure of roads, small airport, limited port facilities adds to disconnection • The plight of minority groups can be extreme
Environmental	• Prone to natural disasters — highly vulnerable people • Lack capacity to adapt to climate change induced droughts and floods • Some environmental degradation — forests for fuel wood and soil erosion • In general environments are less exploited and less pressured and some are in a pristine state (a positive)
Political	• Poor countries low on development scale often have corrupt, non-democratic governments • Resources diverted if country has internal and external conflicts • Limited or no access to trading arrangements/trade blocs

Table 31 Consequences for countries that are bridging the development gap, such as China

Positive impacts	Negative impacts
• Increased investment from FDI and internally • Rising personal incomes from improved range of employment • Improved education and more skilled population • New projects to extend infrastructure, some high profile extend influence of country • Improved international status, membership of WTO etc. • Expansion in core cities should trickle down to the periphery and to the poor • Increasingly 'switched on' by technology of mobile phones/internet	• Over-reliance on export-led growth • Increasing indebtness to international banks for loans • Increasing reliance on TNCs which can dominate decision making • Increasing pressure on services such as waste, health and education • Increasing pressure on land and ecosystems due to expanding development • Increasing pressure on resources for energy • Increasing levels of air and water pollution from industrialisation • Increasing congestion from traffic • Attraction of migrant workers compounds these environmental pressures • Pressure on cultural traditions and values

In contrast to the very poor disadvantaged countries, those countries where economies are beginning to 'take off' experience a different balance of impacts. Table 31 shows that economic impacts are mostly positive but there are many potential negative impacts on the environment. Within these rapidly developing countries there are also several negative impacts on the economy and people.

The impact of the development gap has to be assessed not only between countries but also within countries at a national scale. **Disparities** are apparent between:

- the traditionally poorer rural areas and the urban areas that are experiencing rapid growth as a result of urbanisation and subsequent development
- the core areas (usually ports and large cities) that are experiencing rapid development and the much more stagnant periphery or outback regions that may serve as resource frontiers — the theory is that development will trickle down and spread out to the periphery via a series of stages

Within the megacities that are spearheading the modernisation, the speed and scale of the rural to urban migration that has fuelled such rapid urbanisation has resulted in growing numbers of urban poor. Many lack adequate housing, services and employment. Table 32 shows how development affects urbanisation (initially the urbanisation took place without significant industrialisation). Megacities are frequently a microcosm of the development gap with a sharp divide between a sea of urban poor and the very glitzy centre and a few rich suburbs.

Examiner tip

Look at India and China, the two emergent superpowers, and compare their contrasting development pathways (this links with superpower geographies).

Examiner tip

Familiarise yourself with Friedmann's core periphery theory and Myrdal's concept of growth poles, as they help to explain disparity within a country.

Knowledge check 45

State the reasons for the rapid growth of up to 7% pa for immature megacities.

Table 32 Consequences of the development gap for the world's developing megacities (over 10 million people)

	City type		
	Immature, e.g. Kabul, Lagos, Kinshasha	**Consolidating, e.g. Dhaka, Cairo, Nairobi**	**Maturing, e.g. Mexico City, Sao Paulo, Beijing, Rio de Janeiro, Mumbai**
Economic	Informal street trading (60%), urban farming, small-scale manufacturing represents opportunities; rural to urban migration at 3%+ per year	Increased wealth of people leads to rapid migration continuing, which is difficult to manage in an expanding city (2–3% per year); many jobs in industry and services emerge	Migration rates slow to under 2% per year as the peripheral areas of countryside are developed; job opportunities increase in urban areas, often in professional/high-tech sectors
Social	Dominated by 60%+ slums, especially on urban fringes, squatter settlements for huge numbers of urban poor with no jobs	Increased city wealth, beginnings of planning (waste and water), some upgrading of slums, many have access to basics of health and education	Quality of life satisfactory for many, but for ethnic or social reasons certain groups remain very disadvantaged
Environmental	Major environmental problems from litter, sewage, lack of safe water, disease	Brown Agenda pollution problems from developing industries and growing numbers of vehicles, problems such as rubbish and sewage remain but are improving	Environmental problems being tackled, beginnings of transport, sewage and housing systems; environmental problems are mainly result of growth such as air pollution from cars and factories
Political	Difficult to govern, many protest groups from under classes	Many slum districts such as Kibera, Nairobi, are not officially recognised and are considered illegal by government; a diaspora of tribal and ethnic groups	Government systems increase separation of rich and poor, marked contrasts lead to unrest from 'have nots'; rich live in 'gated' suburbs, occasional riots from under classes

Knowledge check 46

In which group would you put the following megacities?

(a) Bangkok
(b) Manila
(c) Ouagadougou

The development gap has major consequences for people living in many countries. The very poorest people within a country become increasingly marginalised. Their income levels fall significantly below the dominant groups in the same population. This is invariably the result of either overt or covert discrimination, which limits the economic, social and political opportunities available to disadvantaged minority groups. The results can lead to social unrest, outmigration and political protests from which new political movements for justice, freedom or human rights are formed. Figure 30 summarises the issues and reminds you of some interesting case studies that you can research to support your exam answers. Facilitating empowerment of disadvantaged communities is ultimately a priority.

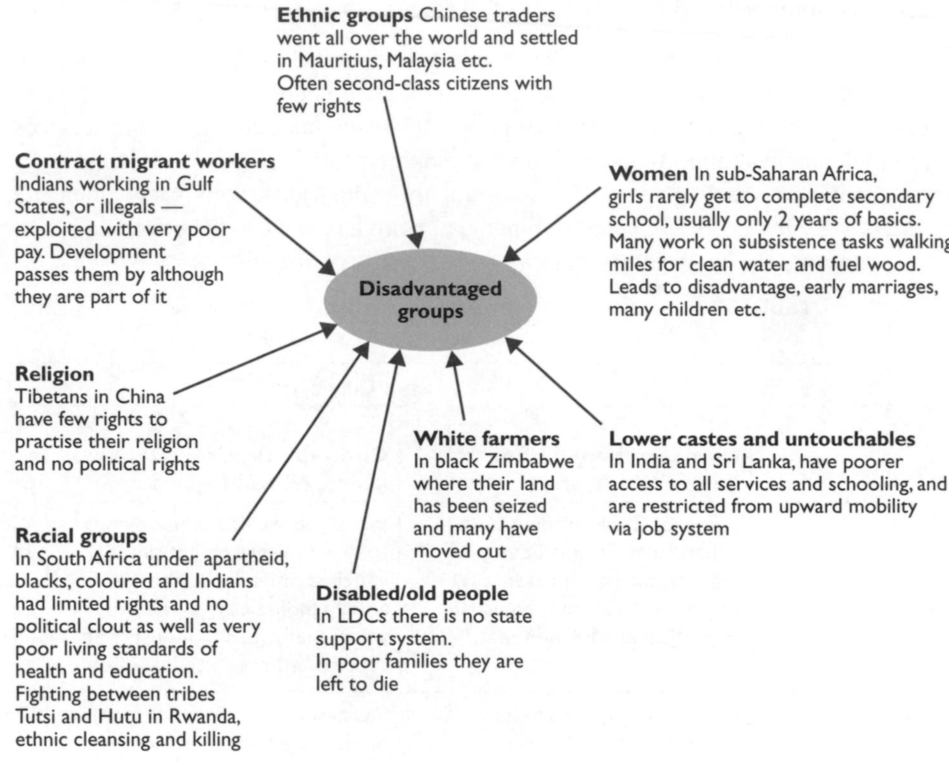

Figure 30 Examples of disadvantaged groups for whom the development gap has had negative consequences

Reducing the development gap

Theoretical underpinning

The theories concerning development (see p. 59, Figure 29) can be used to underpin the different strategies used by governments and organisations to reduce the development gap. These theories are embedded in the ways public, private and voluntary organisations carry out development work and their initiatives reflect the different philosophies.

- **Modernisation theory** (very influential in the 1950s and 1960s) postulates that the underdevelopment of poor countries was largely due to a lack of development of pre-conditions for take-off, i.e. modern socioeconomic structures. Development by TNCs follows this philosophy in emphasising efficient models of production so that developing nations can follow the path of industrialised nations.
- **Neo-liberal economic theory** focuses on free-market economics, intensified by the process of globalisation. Abolishing tariff barriers to encourage trade, privatisation, deregulation and cutting public expenditure are all core to this theory and underpin the actions of the WTO and IMF. Supporters of this theory argue that the emergence of four generations of NICs is proof of the success of the free market. These countries (e.g. China) attract large amounts of FDI.
- The UN utilises aspects of **world systems theory** in attempting to develop networks of global interdependence — for example, in achieving the Millennium Development Goals.
- **Marxist** and **populist** approaches argue for government action to ensure a fairer distribution of wealth both globally and nationally. Traditionally, the Marxist approach was based on a very high centralised communist style economy favouring top-down methods. The populist approach is also left wing, but uses support from people for **grass roots action** using a bottom-up style of development (e.g. Tanzania in the 1980s).
- Non-governmental organisations essentially use a pragmatic approach, which puts sustainable development at the centre with small-scale, localised projects using intermediate technology. This embraces some aspects of **post-development theory**.

Knowledge check 47

Identify the five stages in Rostow's modernisation theory. They form a useful framework for looking at how a country develops economically.

Knowledge check 48

Name one country that currently (2011) follows a Marxist approach.

Knowledge check 49

Distinguish between aid and foreign direct investment (FDI).

Figure 31 summarises the range of approaches available for reducing the development gap. These approaches lead to a number of strategies, which have strengths and weaknesses. Many, such as 'free' trade and structural adjustment plans, are extremely controversial.

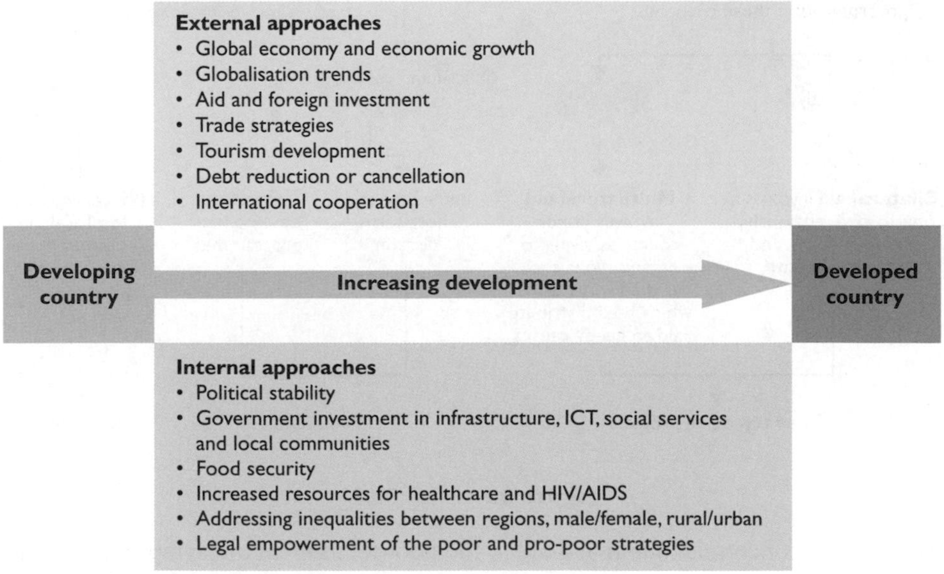

External approaches
- Global economy and economic growth
- Globalisation trends
- Aid and foreign investment
- Trade strategies
- Tourism development
- Debt reduction or cancellation
- International cooperation

Developing country → **Increasing development** → **Developed country**

Internal approaches
- Political stability
- Government investment in infrastructure, ICT, social services and local communities
- Food security
- Increased resources for healthcare and HIV/AIDS
- Addressing inequalities between regions, male/female, rural/urban
- Legal empowerment of the poor and pro-poor strategies

Figure 31 Routes to development

Strategies

Aid

Figure 32 summarises the main types of aid and compares the two models of development — top-down and bottom-up. Overseas aid is the transfer of resources at **non-commercial rates** by one country (donor) or organisation to another (the receiver). This takes the form of not only money but also grants and low-interest loans, goods, food, machinery, weapons and technology, know-how and people such as teachers, nurses and computer technicians. The aim of aid is to help poorer countries develop their economies and to improve services and so raise living standards. The theory seems a good idea, however the reality is more complex. Some types of aid are highly controversial, for example aid **tied** to a particular high-technology project that benefits the donor financially (e.g. military equipment for Tanzania).

While many disadvantaged countries such as Rwanda or Mongolia receive over 25% of their GNP from aid, almost all donor countries contribute less than 1% of their GNP towards aid; most give around 0.4%. Most aid (some 60%) goes to the poorest LDCs, but some middle-income countries that have effective lobbying, such as Israel, receive large quantities and others, such as Turkey (during the Gulf War), receive aid for supporting a war effort.

Donor countries have a pattern of giving that reflects their colonial past or their strategic links; for example, France gives aid to many countries in north and west Africa. China is offering aid to many countries in Africa and is being accused of neo-colonialism as it seeks to secure supplies of resources for its own development.

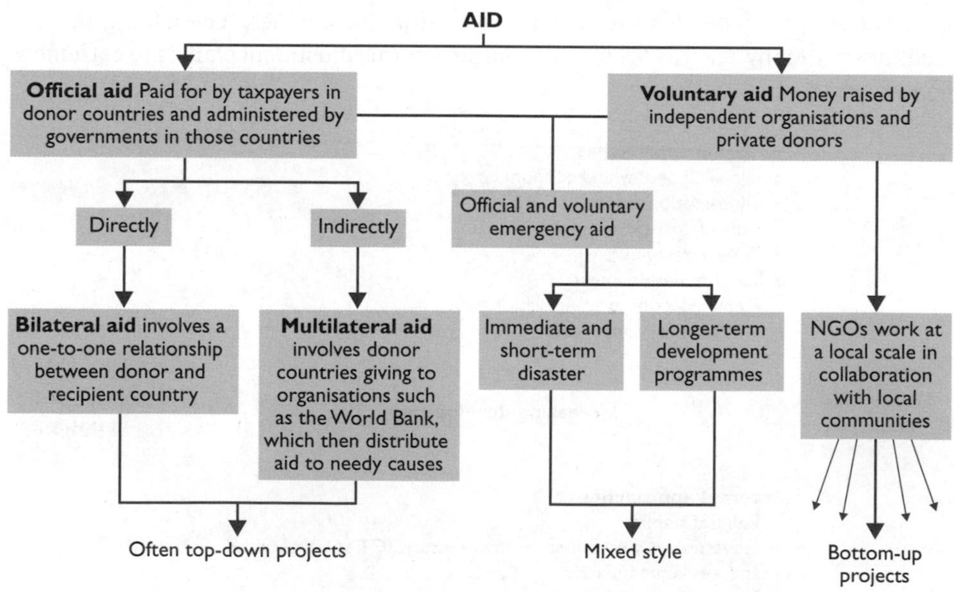

Figure 32 Types of aid

Knowledge check 50

Explain the purpose of emergency aid and how it can be both official and voluntary.

Another key concern is that giving aid to the poor focuses on the symptoms not the causes of poverty. However, few would doubt the value of **emergency aid** after a natural disaster such as the Boxing Day Tsunami in 2004. But even then controversies broke out between donor and recipients as to where, when and how the money should be spent.

Some types of aid are very controversial, such as money for capital-intensive, top-down, mega projects (e.g. World Bank), or loans from the IMF for SAPs (structural adjustment plans), or HIPC programme (eligibility for debt relief) as they impose strict conditions on the recipients that may actually impede development.

Examiner tip

As with many development issues such as trade and aid, there are many sides to the question. Try to explore the controversy via examples to get a balanced view.

Investment

Investment, unlike aid, involves expenditure on a project with the expectation of financial (or social or political) returns. TNCs are the main source of foreign direct investment (FDI) and they invest for profit.

Essentially TNCs from developed countries and NICs invest largely in richer countries as these tend to produce high profits in relatively safe locations. Some of the NICs such as China, and to a lesser extent Brazil, do receive substantial net flows but it is noteworthy that flows to LDCs are negligible. The pattern is also influenced by the type of company — for example, mining companies invest in Australia, Chile or Canada. The incentives offered to investors can influence the pattern — Hong Kong receives over 20% of the 'developing country' total because of the favourable incentives (even more so than in mainland China). So investment is not contributing very much in trying to narrow the development gap between LDCs and the rest.

Knowledge check 51

Which organisations are responsible for
(a) organising global trade
(b) managing countries that get into debt and
(c) lending money to finance multilateral aid projects?

Trade

Trade is possibly even more controversial than aid. At one end of the spectrum are the neo-liberalists who argue that free trade will lift people out of poverty. The vast majority of trade (75%) is between developed countries in North America and western Europe and the newly industrialised countries of the Far East. However, trade does help and can improve the development prospects of many middle-income countries that export value-added manufactured goods to OECD countries.

At the other end of the spectrum are those who argue that trade is responsible for widening the gulf between rich and poor. BINGOs, such as Oxfam, say trade is unfair to developing countries as they do not get fair access to markets because rich countries follow protectionist strategies often via trade blocs. Rich countries control commodity prices, keeping them at a low level, so few developing countries can benefit as much as they should from exports of tin, cotton etc.

Rich countries also dump their subsidised agricultural commodities on international markets at prices below the cost of production. A particular concern is how the patent rules are used by pharmaceutical companies so that (until recently) LDCs could not gain access to affordable medicines.

Knowledge check 52

Explain what a patent rule is.

There have been many negotiations (e.g. Cancun, 2003, and Doha, 2008) to try to make world trade fairer for developing countries. Equally the **Fairtrade** movement (which began in the 1960s) is now a global market worth over £350 million a year. This involves 400+ MEDC companies and an estimated 500,000 small farmers and their families who are usually organised into cooperatives in the world's poorest countries. In this system the producers of food and commodities such as cotton receive a fair price for the products that they grow (above world market price). The problem is that the scale is infinitesimal compared to the overall volume of trade.

Knowledge check 53

Fairtrade is an example of ethical consumption. Explain how.

The future of the development gap — signs of hope

The UN Secretary, Ban Ki-moon, when discussing progress on the Millennium Goals stated that 'Looking ahead to 2015 (the date for the targets to be completed) there is no question that we can achieve the UN's overarching goal of putting an end to poverty. We know what to do, but it requires an unswerving, collective long-term effort'. (This was spoken before the advent of the global depression, see **Note** at the end of this topic, p. 69.)

- There are certainly encouraging signs in terms of trends in world poverty (defined variously as the percentage of people living on less than US$1 or US$2 per day). (In 1990 29%, in 2015 estimated at 10%.)
- In terms of the Millennium Development Goals there is variable progress (see Table 33). However, with the exception of sub-Saharan Africa, all regions are on track to achieve MDG 1 (eradicate extreme poverty and hunger) and huge strides are being made with MDG 2, 3 and 4.

Examiner tip
Use the UN Millennium Goals website progress report to summarise progress by goal and by a range of countries.

Table 33 The Millennium Development Goals

1	Eradicate extreme poverty and hunger	5	Improve maternal health
2	Achieve universal primary education	6	Combat HIV/AIDS, malaria and other diseases
3	Promote gender equality and empower women	7	Ensure environmental sustainability
4	Reduce child mortality	8	Develop a global partnership for development
The website **www.un.org/millenniumgoals/** gives up-to-date information and detail on the MDGs and progress towards them for particular countries			

- Developed countries are as a whole increasing international aid (but usually only to 0.7% of GNP).
- Incremental reform is taking place within the WTO to cut tariffs, trade regulations and quotas to provide developing countries with more access to world trade.
- The World Bank and IMF have reformed their style of aid delivery and means of managing the debts of the very poor countries to try to be more responsive to their needs.
- TNCs are beginning to invest more ethically, more environmentally responsibly and with better working, health and safety conditions in developing countries.
- There have been technological and scientific advances, such as second-generation GM drought-resistant and salt-tolerant crops, which should bring increased food security. ICT supports progress in many LDCs by revolutionising connections to the wider world.
- Many non-governmental organisations (NGOs) have made big differences to poverty by providing appropriate emergency and longer-term aid.
- Several killer diseases have been eradicated and major progress is being made in fighting TB, HIV/AIDS and malaria.
- Increasingly, developing countries are establishing their own local financial institutions (Grameen Bank) and beginning to support other developing countries.

Examiner tip
Select **two** TNCs e.g. Rio Tinto and Shell to study their environmental reports. How much progress do you think they have made?

- There are many examples of successful global partnerships to try to restructure debts and make poverty history — including some progress on climate change.
- Many national governments are increasing their support for deprived minorities (e.g. Aborigines in Australia) in order to empower them.

However, the future can also be considered very depressing for the world's poorest countries, especially those concentrated in sub-Saharan Africa that are experiencing a 'cocktail of catastrophe'. Think about these questions:

- Are we managing to mitigate short-term climate change and adapt to its impacts?
- Are we managing to cope with global depression by reforming global financial systems?
- Are we dealing with global terrorism and corrupt governments?
- For the vulnerable people in the world, are we coping with the humanitarian crises that they face as a result of famine, war and natural disasters (like drought), e.g. in Darfur, Sudan?
- Are we ensuring environmental sustainability for our use of resources (MDG 7) and 'allowing' developing nations socioeconomic sustainability?
- Are the established players really rethinking development and working together to overcome their conflicting priorities?

Examiner tip

Always use the latest statistics and keep up to date, for example by charting progress towards the MDGs or trends in economic growth.

Note: the impact of the 2009 global depression

The credit crunch is hitting the income of the world's poorest peoples the most. It will cost sub-Saharan Africa's people £12.8 billion (UNESCO). This will make the MDGs very difficult to achieve. For example, there is now the prospect of an increase of between 200,000 and 400,000 in child mortality. The IMF suggests that the world's 22 poorest countries might need an additional £20 billion to cope — possibly £100 billion if the depression worsens.

These LDCs have no money of their own to stimulate their economies and progress on reducing poverty (MDG 1) and providing universal primary education (MDG 2) may come to a halt. Both direct and voluntary aid budgets are being squeezed, and developing countries are being offered far lower prices for their exports.

These factors can upset the genuine progress that has been made in reducing the development gap. Even those countries that have powered ahead, such as India and China, are showing signs of recession (around 3% decline in GDP 2008–09).

Synoptic links

The links below show how **bridging the development gap** links to the three synoptic themes (players, actions and futures), and to other units you have studied as well as to wider global issues.

Players

Range from international to local scale. On the one hand many of the international players have partially caused the gap, and on the other hand they are seeking to overcome the consequences. They are involved in developing strategies to reduce it, but their conflicting priorities dominate. New players on the development stage add to this complexity (resurgent Russia, emerging China and India) as they show neo-colonial aspirations, e.g. China in Africa.

Actions

Based on various development theories there is polarisation of opinion about how best to reduce the development gap. In particular, there is a divergence between those favouring large-scale top-down schemes (largely governments and UN agencies) and the NGOs that have used bottom-up strategies effectively.

Examiner tip
Use the Dartmouth Flood Observatory website to locate the floods in 2010. Choose two examples in developing countries and explore how they have led to a humanitarian crisis.

Futures

The actions of the various players will be vital for bridging the development gap. In order to address the plight of the poorest countries in the world (concentrated in sub-Saharan Africa) they will need to act collectively and cooperatively over issues such as meeting the MDGs and making trade fairer. The scale of global environmental, human and economic issues makes their future bleak.

Links to other units

Unit 1

- **World at risk**: the impact of climate change on the continent of Africa and the role of hazards on the vulnerable people in LDCs.

Unit 2

- **Going global**: world cities — the consequences of development and the role of globalisation in leading to 'switched on' and 'switched off' countries in a two-speed world.

Unit 3

- **Energy security**: the way resources (water, energy, biosphere) are shared unequally with developed nations in control.
- **Superpower geographies**: the increasing power of the BRICs and Gulf States supports their ability to bridge the development gap. The latter underpins the techno-fix — with the technology gap very apparent.

Unit 4

- **Life on the margins**: describes the situation of many of the world's poorest people in sub-Saharan Africa.
- **Consuming the rural landscape**: the pleasure periphery has extended to developing nations for whom tourism can be a lifeline to increased wealth.

Summary

- In the twenty-first century development is increasingly seen as a holistic process, including all aspects of human development and not just economic development.
- Development is therefore now measured by a range of indicators, including GDP/GNP per capita and human development index (HDI).
- The development gap is the difference in income and quality of life between the richest and poorest countries of the world.
- Indicators show that the development gap is continuing to widen at its extremities, but it is also narrowing as many countries develop into RICs and NICs.
- There are a whole series of theories used to explain the development gap and the different strategies used by governments. For example, dependency theory (Frank) explains how the poor countries of the South have been exploited by the rich countries of the North, whereas the populist approach emphasises the importance of grassroots development.
- A number of key players operate at a global scale to manage development, debt and trade. Many organisations, such as WTO, World Bank and IMF, aim to bridge the development gap, but at the same time many argue that free trade, top-down development projects and SAPs all actually widen it.
- The development gap has major consequences globally, especially for poverty-stricken, disadvantaged LDCs.
- The development gap is also apparent within countires where poverty-stricken peripheral regions contrast with developed core areas.
- Countries contain many disadvantaged groups.
- Disparity is also very evident in the growing number of megacities.
- Strategies to tackle the development gap include aid, foreign direct investment, trade and debt cancellation, and they all have strengths and weaknesses.
- Some, such as the MDGs, are top-down, contrasting with bottom-up work done by NGOs.
- While there are many signs of hope, progress has been hampered by the global depression.

The technological fix?

The geography of technology

What is technology?

Humans invent technology. It sets us apart from the rest of the animal kingdom. Human inventions include tools, machines and systems. Technology allows humans to control their environment and improve quality of life. Some examples are:

- medicines reducing the impact of bacteria and viruses, prolonging life
- crop breeding increasing yields and thus calorie intake
- just-in-time delivery systems increasing profits

Technology cannot be avoided. People, especially in the developed world, are wholly dependent upon it:

- agricultural technology produces virtually all of our food
- medical technology is relied upon to fix the smallest of ills
- petroleum and nuclear technology supply our energy needs

Examiner tip

Learn a definition of technology for the exam that includes the idea that technology can be tools, machines and systems invented by humans.

The removal of technology, even temporarily, can lead to crisis. In 2007–08, a global crisis was created by oil prices rising to US$147 per barrel. What would we all do if we could no longer afford petrol and diesel for our cars?

While technology is widespread, it is not universal. Table 34 shows how access varies across the development spectrum.

Table 34 Access to technology

	Access to electricity, 2000	Access to improved sanitation, 2004	Internet access population penetration, 2010
Technologies	Power stations; transmission grids	Water supply network; purification equipment; sewage treatment	Computers; wireless or other network
UK	100%	100%	82.5%
Turkey	95%	88%	45%
China	98.6%	44%	31.6%
Bangladesh	20.4%	39%	0.4%

Table 34 also shows how dependency on technology is related to level of development. Access to electricity is ubiquitous in the UK but much rarer in Bangladesh.

Geographical patterns and access

Technology use is strongly related to development. Figure 33 shows the global pattern of internet access in 2007.

Examiner tip
It is important to recognise the complexity of patterns of technology, which are rarely as simple as a North–South divide. Recognising this complexity is a characteristic of A2 geography.

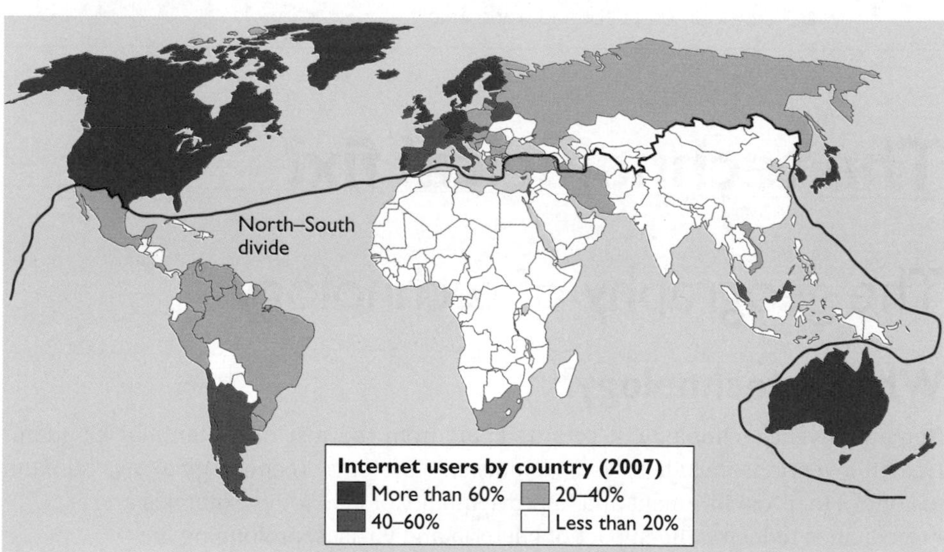

Figure 33 Internet access in 2007

The traditional North–South divide is evident on Figure 33, but the pattern is more complex:

- southern and eastern Europe has fewer users than northern Europe
- Latin America has similar levels to parts of Europe, despite being less developed

- Asia and especially Africa have a low percentage of users, but many Asian megacities such as Mumbai and Shanghai have levels of internet access well over 40%, whereas rural areas close by have levels under 5%.

Economics explains much of the pattern. In Uganda in 2006 it cost US$2,300 for an annual internet connection — far beyond most people's means. There are other barriers to internet access:

- language — most web pages are in English, Chinese or Spanish
- electricity is needed, plus access to a computer
- content needs to be what people want to use
- there needs to be internet service providers

Level of development may explain many differences in the geography of technology, but not all of them.

Air travel has a distinct global geography (Figure 34). Some global regions are very well connected, such as Europe and southeast Asia. Other regions are peripheral. Africa and Russia/Central Asia are only connected in a significant way to Europe.

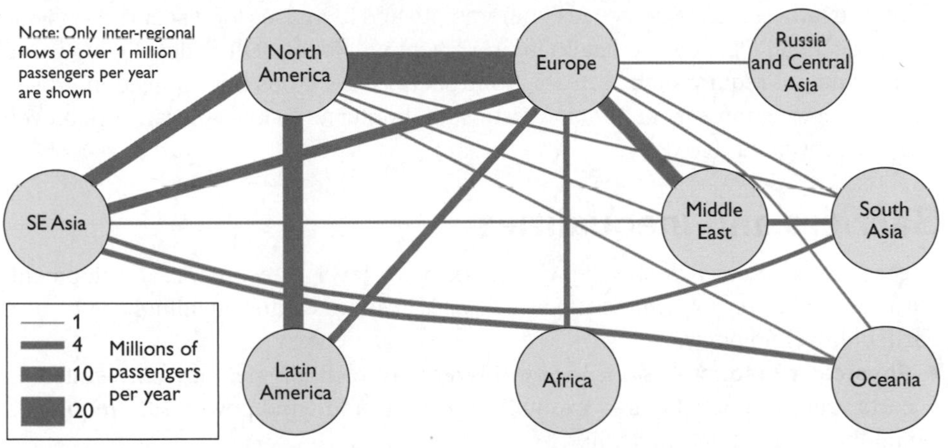

Figure 34 Inter-regional air passenger flows, 2001

Air travel requires infrastructure such as airports and air traffic control systems. Lack of air travel connectivity suggests some global regions are poorly integrated into the global economy and as a result there is little demand for travel. A lack of connections might also mean that new technology takes longer to reach peripheral regions.

Environmental determinism

From time to time we all suffer a bout of environmental determinism — snow prevents travel or a heatwave gives you sunburn. In the developed world this is unusual: sun screen and snow ploughs are readily available. For some people, technology is not available and they are more vulnerable to the environment. Agricultural technology means crop production is now determined much less by the nature of the environment and farmers are less vulnerable to environmental hazards. Table 35 illustrates this. Subsistence farmers in the developing world may have access to almost none of this technology.

Knowledge check 54

Estimate the annual cost of a broadband internet connection in the UK.

Examiner tip

The internet is a fast-changing technology. Keep up to date with the latest global patterns at **www.internetworldstats.com/stats.htm.**

Examiner tip

While access to technology is strongly related to level of development, be careful not to claim that it is all about how much money people have. Access to technology is controlled by a range of factors.

Knowledge check 55

What are 'subsistence farmers'?

Table 35 Agricultural technology and vulnerability

Farm technology	Role	Vulnerability without the technology
Irrigation	To provide additional water for crops during dry periods, or all of a crop's water in dry regions	Dry regions produce fewer crops in a shorter season; drought can lead to crop failure
Pesticides	Sprayed on crops to kill pests and increase yields by decreasing crop losses	Expected yields factor in a % loss to pests; crops are vulnerable to pest plagues
Fertilisers	Added to soil to provide additional nutrients for growth — to be effective often requires irrigation	Yields are restricted by the natural nutrients available in the soil
Farm machinery	Used to replace human labour and increase efficiency and farmed area	Size of farmed area is determined by population and distance they can travel; crops may have to be abandoned during floods or severe weather due to lack of manpower
Hybridisation	Inter-breeding of crop varieties under controlled conditions to produce disease or pest resistance, and higher yields	Crops could become vulnerable to a pest or disease with no viable replacement

Access to farm technology reduces vulnerability and increases food security. If yields can be secured, lifestyle and health improvement will follow. Often developing world farmers do not require high-tech solutions such as GM crops but do need improved seed varieties, simple irrigation and fertilisers. In much of sub-Saharan Africa even this has proved hard to deliver.

Examiner tip
Consider some of the environmental costs that using the farm technology in Table 35 could result in. Remember that all technology is a trade off between costs and benefits.

Barriers and inequalities

Access to technology is strongly correlated with level of economic development, but there are other explanations for lack of access to, or an unwillingness to use, particular technologies.

- **Physical reasons** — some renewable energy technologies are only suited to certain physical locations — examples are solar and wind power. HEP requires a suitable water supply and valley.
- **Political reasons** — in North Korea internet access is not available to ordinary citizens. In order to control the flow of information that people receive and also to ensure the 'correct' political message is maintained, the government prevents internet use.
- **Environmental reasons** — certain groups voluntarily shun certain technologies. Organic farmers do not use pesticides or cattle antibiotics because of their supposed negative environmental and health impacts.
- **Religious reasons** — contraceptive technology is rejected by some religions, such as the Roman Catholic Church; other religions, such as Islam, accept some forms of contraception but not others.
- **Military reasons** — nuclear technology has been controlled by the international Nuclear Non-proliferation Treaty, which aims to prevent nuclear weapons falling into the 'wrong' hands. The International Atomic Energy Agency tries to ensure states with nuclear power do not use it to develop nuclear weapons.

All people have views on technology. Parents might consider the Xbox as 'dangerous' to children's health. Many people are ethically uncomfortable with genetic engineering. People who reject technologies are referred to as technophobes.

Examiner tip
Learn a range of examples of where and why technology is not available to some people.

Humans invent technology but simply inventing a technology does not make it available. New technology is often expensive. Inventors and developers invest time and money in invention and the law ensures they get a return on their investment:

- New inventions are protected by intellectual property rights.
- A patent is given to the inventor of a new technology protecting it from being copied.
- Inventors usually license companies to manufacture the new technology, and receive financial royalties.

Royalties and licence fees keep new technologies expensive until patents expire (usually after 20 years). The patent system could prevent a new drug from being made widely available because the pharmaceutical company that invented it charges a high price to recoup its R&D (research and development) costs. For this reason Thailand and Brazil have begun producing generic drugs (illegal copies) to treat AIDS/HIV patients and have lowered monthly treatment costs from US$500 to US$30 per patient.

Knowledge check 56

What would a drug company's view probably be of developing countries that produce cheap, generic copies of patented drugs?

Technology and development

The technology gap

Developed world economies are increasingly knowledge based as we move from the industrial age to the information age. In a knowledge economy, ideas, information and services make money, not goods. The growth of the knowledge economy has been promoted by:

- globalisation of markets and free trade
- information and communications technology
- networking using internet technology
- high-tech products and services

Much of the developing world is still industrial. Apple's iconic iPhone illustrates this technology gap (Figure 35). As with many industrial products, complex components are made in Japan and the NICs. Less complex assembly is completed in China. The least developed world, e.g. Africa, plays no role in iPhone design or manufacture.

Knowledge check 57

What barriers might prevent TNCs setting up factories to produce iPhone components in African countries?

Figure 35 The Apple iPhone

R&D spending is one explanation for this:

- globally about US$1 trillion is spent every year on R&D
- one third of this is spent in the USA
- in the developed world there are 2,000–5,000 research personnel for every million people but only 10–50 in the least developed countries

R&D allows the technologically developed to stay that way. Most technological innovations, as measured by patents granted, originate in the developed world. In 2007, 51% of all new world patents originated from the USA, 20% from Japan and 16% from the EU. Patents gain royalties and licence fees and these show an extraordinary concentration in the rich world (Figure 36).

Developed world governments and TNCs invest huge sums in R&D, and high rates of funding for universities ensure a steady stream of skilled graduates who can carry out yet more R&D.

Examiner tip

Learn some key facts and figures relating to R&D, patents and royalties as these tend to be heavily concentrated in the developed world and in many ways are a barrier to the wider adoption of some new technologies.

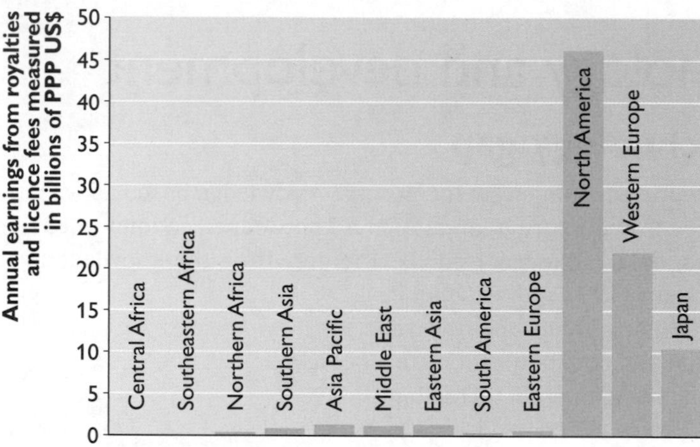

Figure 36 Royalties and licence fees, 2004

Leapfrogging

The developing world would like nothing more than to break into the R&D, patents and royalties club. However, this is very hard to achieve:

- skilled researchers are required, meaning investment in universities
- much R&D spending is by TNCs — there are few in the developing world
- governments' budgets are earmarked for water and housing projects; there is little spare money
- laboratories and research facilities are expensive

NICs such as Taiwan and South Korea have become big R&D spenders, but only slowly. Samsung, a South Korean TNC, spent more on R&D in 2007 than IBM.

Table 36 Comparing two leapfrogging technologies

Mobile telephones	Technology	Lifestraw
Communication	**Use**	Water purification
A telephone landline network	**Leapfrogged**	A water purification and distribution network
Requires installation of a mast network, and access to electricity to charge phones; villages often have 'mobile' chargers linked to a car battery; limited by signal coverage	**Flexibility**	30 cm long tube (straw), which purifies water using filters, iodine-coated beads and active carbon; can be used anywhere and lasts about 1 year
Around US$40 per year	**Cost**	US$3 per person per year
Increases ability to search for jobs, keep in touch with family, access prices at markets; can be used to warn of natural hazards	**Impacts**	Major improvements in health — kills virtually all bacteria and parasites; does raise iodine levels in users, although many are iodine deficient anyway

For other developing countries, technological leapfrogging provides one possibility. This is when a technology is adopted without a precursor technology. The classic example is mobile phones, which have been adopted in countries that have never had an extensive landline network. There are other examples:

- laptops and WiFi — without a hardwired network stage
- solar panels and micro HEP — without a complex electricity transmission grid

Technologies that can leapfrog are usually mobile and physically small. They also need to be stand alone — cars, for instance, must have roads. Table 36 compares two leapfrogging technologies.

One criticism of mobile phone technology, lifestraw, laptops and GM crops is their developed world origins. This means developed world companies collect royalties and profits from the developing world — the very people who need technology the most, but can least afford it.

Examiner tip

You need to have a range of examples of leapfrogging for the exam, not just the rapid spread of mobile phones in the developing world. One example is never enough!

Knowledge check 58

What type of technologies are most often involved in leapfrogging?

Costs and benefits

One view of technology is that 'technology is neither good, bad nor neutral' (Kranzberg's first law of technology). This means that technology always has some impact. Most technologies are introduced with a particular aim but may have other, unforeseen impacts. These impacts are referred to as externalities. They are costs or benefits that are not accounted for in the financial cost of a product. The introduction of the Green Revolution in the 1960s and genetically modified crops in the 1990s both illustrate this (Table 37).

The Green Revolution increased food supply, especially in Asia. Since the 1960s, HYVs have had to be developed every few years to replace those that succumb to pests and disease. The Gene Revolution is more problematic. Currently, much of the production is exported as fibre (cotton) or for cattle feed (maize and soy), so food security has often not increased, although with GM2 this may improve. Both technologies have had unforeseen environmental impacts and often led to social polarisation.

Examiner tip

Learn definitions of both Green Revolution and genetically modified farming, as they are often confused in the exam.

Table 37 Developing world Green Revolution and Gene Revolution costs and benefits

Technology and original aim	Economic impacts	Social impacts	Environmental Impacts
Green Revolution High yielding crop varieties (HYVs) plus fertilisers, irrigation and machinery double or treble wheat and rice yields, increasing food security in the developing world	+In many cases yields increased dramatically +Two crops per year could be grown and harvested +Increased food security	–Introduction of machinery leads to unemployment and increases rural–urban migration –Only relatively well-off farmers can afford the new technology +Improved diet and health	–Increased use of fertilisers causes nutrient rich runoff and eutrophication –Pesticide over-spray damages biodiversity –Some HYV monocultures wiped out by pests and disease
Gene Revolution Genetic make-up of crops (maize, cotton, soy) altered so they are resistant to pests, disease and herbicides, or tolerant of drought; yields and food or income security increase	–Farmers become dependent on seeds and chemicals from TNCs such as Monsanto –Some studies suggest yields have not increased +Increased exports and rising farm incomes	–Public opinion, in countries such as the UK, reject the technology –In Argentina, larger GM maize farmers have tended to buy out smaller ones, leading to social polarisation –Many of the crops are for export, not food	–Weeds may be developing resistance to herbicides –Deforestation in Latin America in order to increase the farmed area

Dealing with externalities

Using technology has consequences. Generally speaking the more technology we use, the greater the environmental impact. This is because:

- goods require resources, which have to be extracted and processed
- manufacturing causes pollution
- technology has to be powered or fuelled, which usually means fossil fuel use

It is no surprise to find that the most technologically advanced societies use the most energy resources and have the largest ecological footprint (Figure 37).

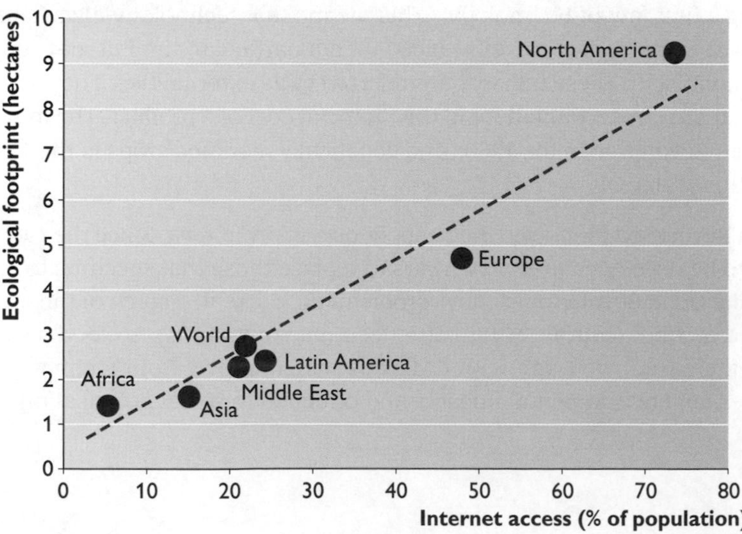

Figure 37 The relationship between internet access and ecological footprint

Extensive use of technology leads to carbon dioxide emissions and global warming, with large potential impacts for human and ecosystem wellbeing:

- rising sea levels may flood coastal cities
- increased sea temperatures may cause coral bleaching
- increased drought may lead to crop failure and water shortages
- increased flooding may destroy homes and livelihoods

There is increasing pressure to move towards a 'polluter pays' model. This would mean that if people treat the environment as a pollution sink, they have to pay. The most obvious way to do this is with 'green taxation':

- the EU Emissions Trading Scheme sets pollution quotas — exceed the quota and carbon credits must be bought
- UK car tax is now linked to carbon dioxide produced per km

Knowledge check 59

Explain what is meant by 'polluter pays'.

These taxes are designed to encourage pollution reductions. In reality, it is the type and the energy intensity of our technology that causes global warming. If the technology we used was less demanding on resources and more renewable, the impacts would be much smaller.

Technology, environment and the future

Contrasting approaches

Most inventions aim to make life better — new drugs, labour-saving machines or improved crops. Unfortunately, humans produce weapons too, so not all technology is 'good'. Some technologies are considered to have such negative consequences that attempts are made to ban or control them — landmines, chemical weapons, CFCs and genetic modification fall into this category.

Examiner tip

A useful example of a 'good technology gone bad' is the pesticide DDT. You can explore this example at: **www.chem.ox.ac.uk/it_lectures/chemistry/mom/ddt/ddt.html**.

Technology is used as part of the development process. Table 38 shows there are philosophical differences between what is considered the 'best' type of technology to use.

Examiner tip

There are many different classes of technology (high-tech, intermediate, appropriate, renewable etc.), which often overlap. Make sure you learn definitions for each and have examples for all.

Those in favour of intermediate (or appropriate) technology argue that in the developing world technology needs to be fit for purpose. This means low cost and easy to repair locally. High-tech solutions mean people rely on high-tech companies. Intermediate technology is often environmentally friendly, as is alternative technology. The difference is that alternative technology can be high tech — such as a hydrogen-fuelled car.

Intermediate technology is often bottom-up. Local people and organisations are involved in planning, building and maintaining the technology, and take ownership of it. Large engineering schemes and high-tech projects are often controlled by governments and TNCs. Critics argue that this means local people's needs are often not met.

Examiner tip

Research a top-down mega project with a contrasting intermediate technology bottom-up project.

Table 38 Which type of technological fix?

'Small is beautiful'	'The bigger the better'	'High tech is best'	'Renewable future'
Intermediate technology — low cost, simple, small scale, using local skills and resources	Mega-engineering projects that provide a one-off solution at very high capital cost	Most advanced solution currently available, e.g. nanotechnology, bioengineering and electronics	Alternative technology — lowest possible environmental impact and pollution
Example: village hand pump installed by an NGO	Example: large dam funded by government	Example: nanofiltration systems from TNC R&D labs	Example: solar powered water pump — TNC/NGO joint venture

The big fix?

Perhaps the ultimate technological fix would be technology that could reverse the global environmental problems that humans have caused, for instance global warming and land degradation. These fixes have been researched and are referred to as planetary or geo-engineering.

- Proponents of geo-engineering argue that techno-fixes for global warming are more likely to work than persuading people to change their lifestyles (an attitudinal fix) to reduce pollution.
- Opponents of geo-engineering argue that such solutions are effectively a big experiment with unknown outcomes.

Examiner tip

Geo-engineering refers to regional- or global-scale technology that re-engineers the way the planet works. It is often transboundary, so if one country implemented it this could lead to conflict with other nations opposed to it.

Arguing against technology on the basis of possible unknown outcomes is referred to as applying the 'precautionary principle'. Those against GM crops often argue that GM material might accidentally be transferred from crops to wild plants, with unknown consequences. Because it has not been proved that this can't happen, GM crops should not be used. Table 39 examines past and future techno-fixes.

Table 39 Past and future techno-fixes

PAST FAILURE **The Aral Sea disaster**		**FUTURE POSSIBILITY** **Artificial global dimming**
Transform arid USSR steppe lands into productive cotton-growing farmland using mega-scale irrigation	**Aim**	Reduce solar input by creating an artificial aerosol blanket in the atmosphere to reflect sunlight back into space
Dams and diversions of the Amu Darya and Syr Darya rivers redirect river flow along 40,000 km of canals to irrigate an extra 3.5 million hectares of land	**Technology**	Using aircraft, rockets, artillery or balloons to 'shoot' sulphur dioxide aerosols into the stratosphere to increase concentrations
• Syr Darya and Amu Darya rivers dry up • Aral Sea shrinks to 25% of its original size • The exposed sea bed is scoured by wind creating 'salt storms' dumping salt and farm chemicals on people and land • Chemical pollution increases local incidence of cancers • Fishing industry collapses • Flora and fauna in the sea have died out	**Impacts and problems**	• Aerosols cool the planet by reflecting incoming solar radiation and promoting cloud formation • Calculating the amount of sulphur dioxide required to create the desired cooling could be difficult • Additional sulphur dioxide could increase the acid rain problem • Cooling could have wider consequences and knock-on effects altering climate in unforeseen ways

The Aral Sea is often cited as the world's worst ecological disaster — increasing cotton production but destroying the ecology of the world's fourth largest inland sea.

Opponents of geo-engineering argue that we can never know the full impacts of such large-scale schemes until it is potentially too late. Any further attempts to solve global environmental problems with similar schemes will need to be progressed with extreme care.

Knowledge check 60

What would many environmentalists think about the artificial global dimming technology outlined in Table 39?

Technology and sustainability

Countries have very different attitudes towards technology and its environmental impact. It is generally accepted that for humans to have a high quality of life they need to live in a healthy environment. This is the 'egg of wellbeing' idea shown in Figure 38.

Country	GDP per capita US$ (PPP)	*The Economist* quality of life index	Environmental sustainability index	Comment on ecosystem versus human wellbeing
Sweden	30,600	7.9	71.7	Both high, balanced
Uruguay	8,900	6.4	7.8	Both relatively high, despite middle income
Sri Lanka	3,800	6.4	48.5	Human wellbeing exceeds ecosystem wellbeing
USA	41,500	7.6	52.9	Low ecosystem wellbeing, high quality of life
Kuwait	14,550	6.1	36.6	Low ecosystem wellbeing despite higher income
Ghana	2,600	5.2	52.8	Ecosystem wellbeing similar to USA, low quality of life

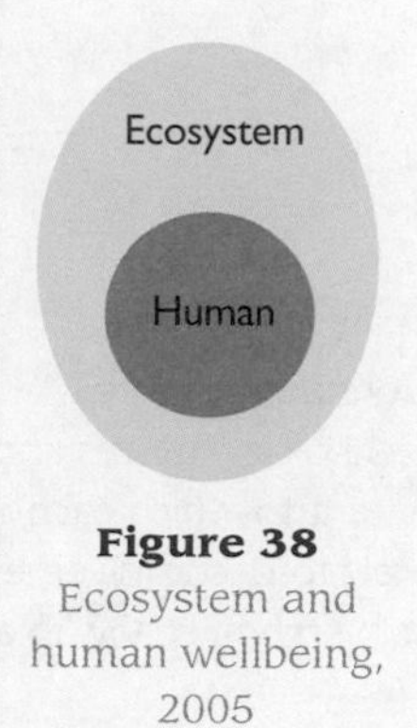

Figure 38 Ecosystem and human wellbeing, 2005

Figure 38 shows that Sweden manages to maintain high ecosystem wellbeing as measured by the environmental sustainability index (ESI), and high human wellbeing as measured by GDP and *The Economist* quality of life index. This is a rare balance. In the USA, ESI levels are similar to Ghana, despite the huge difference in wealth. In Kuwait, wealth seems to come at the expense of the environment. Uruguay manages a good human and ecosystem balance even though it is not rich.

Technology and environmental sustainability are not mutually exclusive. However, some hard choices have to be made if the use of technology is to be compatible with the concept of sustainability (Figure 39).

Examiner tip

Make sure you learn a definition of environmental sustainability for the exam and can apply it to different examples of technology to judge their sustainability.

- Technology must be cost effective and affordable, for instance not leading to debt
- Externalities must not pass costs on to others

- Technology should produce little pollution
- It must not have adverse consequences for ecosystems

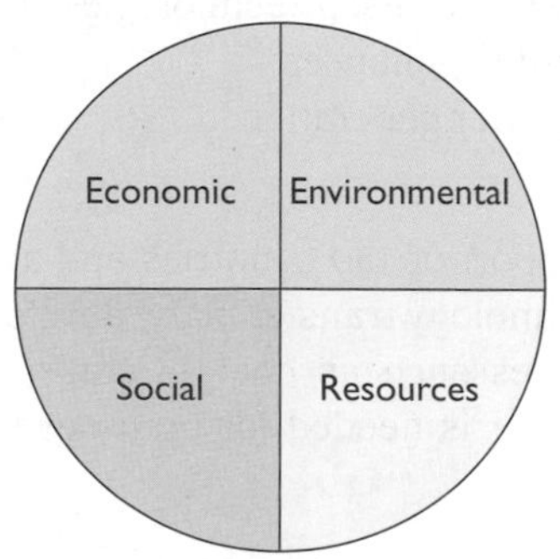

- Technology should benefit all parts of society and not polarise it
- Technology should promote human health and wellbeing

- Technology should use renewable resources
- Power for technology should avoid fossil fuel use

Figure 39 Technology and sustainability

Knowledge check 61

Explain why wind power is more environmentally sustainable than using coal.

Some technologies fit the criteria in Figure 39 better than others, for instance, wind and solar power rather than coal or gas. Current technologies such as the motor car, and large-scale technological fixes such as mega dams, do not fit the criteria as well. This might suggest some difficult decisions will have to be made in the future.

Futures

The future of our technological world is not clear. There are several possible futures.

- A divergent world with technological advances in the developed world, but lack of access to technology in the developing world. Divergence would ensure large numbers of people were on the wrong side of the technology and development gap. Many would continue to rely on aid, especially in the aftermath of major natural disasters.
- A convergent world with increased transfer of technology to the developing world. Convergence would begin to bridge the technology gap. However, it would require a major technology transfer that the current system of patents and licensing works against. Increased technology use in the developing world could intensify global warming if energy sources follow a business as usual model.
- A world that switches towards renewable resources for making and powering technology. Technology powered by oil, coal and gas may lead to environmental disaster and economic uncertainty as fossil fuels run out. Switching to renewable technologies would help avoid both negative outcomes.

Examiner tip

Consider the three technology futures and the extent to which each is likely to occur (divergent, convergent, renewable).

There is some evidence of convergence, albeit in only a few countries. China is approaching Japan in R&D spending, which grew by 23% between 2001 and 2006 compared to only 1–2% in the USA. In South Korea, companies spend about 6.5% of their budgets on R&D compared to only 5% in Europe. These countries are NICs and have gradually built their wealth. Many now have governments with deep enough pockets to spend heavily on innovation and also have their own TNCs that invest in R&D. In India, some home-grown TNCs have bought slices of high-tech industries such as Tata's buy-up of Jaguar Land Rover and Mittal Steel's 2006 purchase of Luxembourg-based Arcelor.

Examiner tip

Technology transfer is an important concept as it could provide a way for developing countries to gain access to the technology they need to improve human and ecological wellbeing.

In the least developed world technology transfer currently relies heavily on the development work of NGOs and bodies such as the UK government's Department for International Development. There are signs that this may change. In 2008, the Global Environment Facility (GEF) was given responsibility for technology transfer to the developing world. GEF works to transfer technologies such as:

- energy efficient lighting and appliances
- efficient and renewable power generation
- fuel-cell buses

The GEF, which has the support of 180 countries and a US$3 billion annual budget, recognises that without technology transfer many developing countries will continue to use polluting technologies such as coal — contributing even more to global warming. Technology transfer is needed for both development and environmental sustainability.

Synoptic links

The links below show how **the technological fix?** links to the three synoptic themes (players, actions and futures), and to other units you have studied as well as to wider global issues.

Players

Much of the most advanced technology is developed by global TNCs that research innovations and bring them to the market place — this includes technologies such as GM crops, new drugs and IT applications. Technology is rapidly adopted in the developed world. In the developing world NGOs are often responsible for introducing appropriate and intermediate technology to people who badly need it.

Actions

Big, high-tech fixes tend to be favoured by governments as they provide visible solutions to national problems — the Three Gorges Dam is a classic example. Intermediate and appropriate technology, which is usually smaller scale and often local, is favoured by those who believe in grass roots approaches and sustainable development.

Futures

The technological future could be business as usual. This would involve the developed world continuing to adopt new technology, with access to it in the developing world being limited. A more sustainable future might involve wholesale technology transfer from the developed to the developing world — in order to improve quality of life. This could be enhanced by the more radical adoption of renewable technology to replace energy intensive, polluting technologies in transport and energy supply. Geo-engineering solutions fall into the radical category, but are highly controversial 'big fixes'.

Links to other units

Unit 1

- **World at risk**: technology may play a role in managing the threat of global warming, from adaptation strategies such as coastal defences, to large-scale geo-engineering technological fixes.

Unit 3

- **Water conflicts**: high-tech engineering solutions such as large dams, water transfer schemes and desalinisation, as well as intermediate technology, are used to try to match water supply and demand.
- **Energy security**: renewable energy technology is important in securing energy futures.

- **Bridging the development gap**: there are contrasts between the often high-tech, top-down style of development project and the intermediate technology, bottom-up approach.
- **Superpower geographies**: the rich and powerful develop most new technology by investing in research and development; they often control access to technology through patents, royalties and licence fee payments.

Unit 4

- **Tectonic activity and hazards**: technology plays an important role in hazard management from monitoring and prediction, to relief and reconstruction.

Links to wider global issues

- The **global environmental crisis** includes issues such as global warming, loss of biodiversity, water supply degradation and soil erosion. Technology can help control or even solve many of these problems. Often the key is to make sure technology is applied appropriately so that it is affordable and manageable. It should not replace one problem with another by, for instance, using excessive amounts of energy.
- Many people in the world find themselves on the wrong side of the technological **development gap**. If the gap is to be bridged it is important that there is improved access to communications, energy supplies, drugs and farm technology, which will lead to improved food security, a decrease in poverty and a reduction in vulnerability.

Summary

- Technology allows humans to control their environment more effectively and improve their quality of life.
- Technology encompasses a growing number of branches, ranging from mega high-tech projects to intermediate small-scale, low-cost technology. New branches are emerging from nanotechnology to biotechnology to ICT.
- Access to technology varies across the development spectrum. The pattern is more complex than a straight North–South divide of haves and have-nots, as a result of socio-cultural and political factors as well as those linked to economic development.
- Barriers to the development and use of technology are physical, environmental, political, cultural/religious as well as economic, making the patterns of access very uneven.
- While a technology gap exists, broadly reflecting the development gap, leapfrogging by developing countries can narrow it.
- Leapfrogging involves the adaptation of advanced technology without precursor technology such as landlines, wires and cables.
- Technologies such as the Green Revolution or GM have both costs and benefits as their environmental, social and economic impacts can be both positive and negative.
- Technology needs to be appropriate for the chosen purpose and situation, for example desalination plants versus rainwater harvesting.
- Some technologies are controversial, especially when applied to solving the world's global environmental problems, for example the use of geo-engineering to combat global warming, should the attitudinal fix of mitigation fail.
- While the techno-fix can be sustainable, many technologies contribute to an increase in eco-footprints.
- The future of technology in the world is difficult to predict. A divergent world with developing companies relying on ongoing polluting technology, or a convergent world where developed nations transfer technology to narrow the development gap? A green sustainable technological world, or a world where technology increases the dangers to the world's peoples?

Questions & Answers

Assessment

The Unit 3 examination consists of a 90-mark, $2\frac{1}{2}$-hour exam, split into Section A and Section B.

Unit 3 Contested planet 60% of A2 marks	**Section A** • Two 25-mark resource-based longer essay style questions from a choice of five. • Each question will be split into part (a) worth 10 marks and part (b) worth 15 marks.	50 marks (80 minutes)
	Section B • A synoptic issues analysis taking the form of three, linked essay style questions. • 4 working weeks prior to the exam a synoptic resource booklet will be given to you to work on in class.	40 marks (70 minutes)

Section A tests the depth of your knowledge and understanding, but you can choose the two questions that you want to answer from the five provided. You will have around 40 minutes to do each Section A question. Section A questions are in two parts:

(a) Data stimulus (10 marks)

(b) Extended writing (15 marks)

All of the questions in Sections A and B are marked using levels mark schemes, and the quality of your written communication is important.

Synopticity and the Section B issues analysis

Unit 3 is the synoptic unit. In geography, synoptic assessment tests your ability to:

- make links between different topics within Unit 3
- see wider links between Unit 3 and the other AS and A2 units
- apply themes, models and concepts to structure your work
- draw parallels and contrasts between different examples and case studies

The specification highlights three themes that cut across all of the topics in Unit 3. These three themes are players, actions and futures.

- The views of different **players** and their role in both creating problems and managing solutions.
- The range of **actions** that could be used to try to solve problems and implement solutions, and different scales of action from local and personal, up to global and international.
- The range of '**futures**' that humankind might aim for, such as a business as usual future much like today, a more sustainable future in terms of development and environment, or a more radical future based on green growth.

The players, actions and futures themes of Unit 3 are synoptic concepts. They cut across all of the topics in this unit and will be assessed as part of the synoptic assessment in

Section B of the exam paper. In each exam series one of the six Unit 3 topics will form the basis of the synoptic issues analysis pre-release resources.

You will find out which of the six topics is the basis for the Section B issues analysis when you receive your pre-release resources. The selection of topic for each exam series is entirely random.

In Section B there will be a sequence of three questions linked to the issues analysis resources. These questions are worth 40 marks. In general they will follow a sequence similar to this:

Question 6a

An examination of the issue, explaining the issue and perhaps examining a range of viewpoints, or picking out key factors or causes.

To some extent this will be a 'background' question to help you get to grips with the topic and explore different viewpoints.

Question 6b

An evaluation of the options or impacts.

This could be an in-depth question where you examine and use evidence from the resources provided, your own research and other parts of your geography course.

Question 6c

An assessment of solutions or a justification of a decision.

This question could be in the form of a detailed conclusion or an overview. It will tend to require you to carefully sum up and use evidence to support your view.

Top tips for Section B

You need to think carefully about how you deal with the pre-release issues analysis resources in the run-up to your exam. The resources will consist of a 5-page booklet with a mixture of text, graphics (maps, graphs, tables), opinions or views and websites.

Before the exam:

- Make a glossary of all the terminology used in the pre-release resources, so you understand its meaning.
- Make a series of mind maps showing how the pre-release resources link to:
  - the specification content for the chosen topic, e.g. energy security
  - other topics within Unit 3
  - topics covered in AS Units 1 and 2, and A2 Unit 4
- Consider links to wider, over-arching themes such as climate change and sustainability.
- Research, using the websites provided. Ideally make a set of research notes rather than printing large volumes of web material.
- Analyse the opinions and views.
- Consider the broad themes that questions could focus on, but avoid trying to 'question spot'.
- Use your teachers. Ask them to explain any areas you are unsure about, or organise some group activities in class to help you understand the resources.
- Look at past papers and sample assessment materials and think about the sequence of questions, and how they link to each other.
- Consider researching some parallel and contrasting examples if you don't have these in your Unit 3 file.

In the exam:

- Read the questions and think carefully about the command words. These are likely to be 'assess' and 'evaluate', or similar command words, and they demand higher level skills.
- Use the resources, and quote them. Try to bring in your own supporting examples and facts and figures.
- Watch the clock, to make sure you finish all Section B questions.

Question 1 **Energy security**

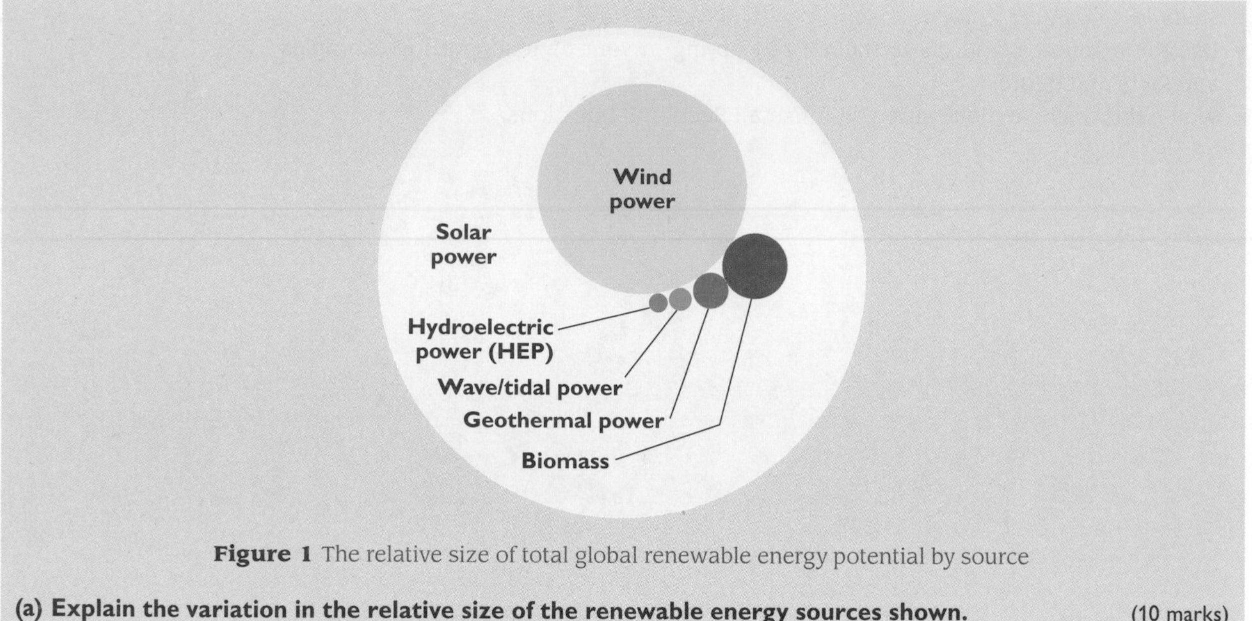

Figure 1 The relative size of total global renewable energy potential by source

(a) Explain the variation in the relative size of the renewable energy sources shown. (10 marks)

ⓔ Focus your answer on giving reasons for the different sizes of the circles in Figure 1, and try to mention all energy sources shown. Note that the data are for potential, not current, energy use.

Mark scheme and levels

ⓔ The variation in renewable potential is large, from a small amount of HEP to a very large solar potential; solar and wind power dwarf the other four sources shown. Solar power is available almost anywhere on Earth; its potential is reduced by cloud cover and latitude (e.g. Arctic/ Antarctic). Wind power has large potential because it can be harnessed wherever the wind blows; there is greater potential out at sea, and some areas (e.g. the UK) are windier than others. HEP is restricted by the availability of water and a suitable reservoir site; it has the most physical restrictions on its development. Tidal power needs a usable tidal range such as the Severn Estuary. Geothermal power needs hot rocks relatively close to the surface in tectonically active areas such as Iceland — not available everywhere. Biomass only requires land to grow crops, explaining its larger availability and more global distribution — however, land is needed for food, limiting its potential.

Level	Mark	Descriptor
Level 3	8–10	Detailed explanations across the range of renewable sources with some use of supporting examples and places. Explanations are always clear. Geographical terminology is used with accuracy. Grammar, punctuation and spelling errors are rare.
Level 2	5–7	Some explanations for some of the renewable sources but unbalanced. Explanations are clear, but there are areas of less clarity. Lacks full range. Geographical terminology is used with some accuracy. There are some grammar, punctuation and spelling errors.

Level	Mark	Descriptor
Level 1	1–4	Descriptive use of Figure 1 with some suggestions about size variation but these are very generalised. Explanations are oversimplified and lack clarity. Geographical terminology is rarely used with accuracy. There are frequent grammar, punctuation and spelling errors.

Student answer

The potential of renewable energy sources is determined largely by physical factors. However, improvements in technology can sometimes mean that physical factors can be expanded, for example second-generation biomass fuels such as Jatropha can be grown on poor land, and in the case of third-generation biofuels, large ponds are needed for the algae to breed — so the issue of land becomes less contentious than it is for bioethanol and biodiesel (first-generation fuels). **a**

Solar power is the largest circle on the diagram — basically, with improved technology there is potential for certain types of energy such as photovoltaics across most of the planet. Some places such as the Sahara desert (an ideal location) are being seen by countries such as Spain as having potential for outsourcing power installations. **b**

Wind power also has widespread potential, especially in the westerly wind belts such as the Atlantic and the North Sea for the UK or 'Roaring Forties'. Again, improving wind turbine technology combined with more efficient storage means that many more locations could become viable. Wind farms for Denmark, or the UK, or California USA have huge potential to make a significant contribution to their countries' energy security. The improved technology for turbine anchorage has opened up many deeper-water offshore areas, which is probably where the future lies. **c**

Biomass is a controversial potential source, as currently biofuels come from crops such as soya beans, sugar cane, palm oil, maize — these crops for fuel compete with the same land that produces food crops, and also can involve acres of forest being cut down — so far this is why the potential of biofuels is currently limited. **d**

The other three sources are more localised and therefore have less potential — for example, geothermal power is linked to the decay of hot rocks as in Iceland, where it forms a highly significant local source. Tidal power is currently restricted to certain estuaries, often of outstanding ecological value, which have a strong tidal range, such as the Bay of Fundy or the Severn Estuary. HEP's low energy potential is perhaps a little surprising, but one explanation may be that many of the most suitable sites have been used up, with many potential sites being highly controversial, such as those in National Parks **e**. It is possible too that climate change has the potential to diminish the reliability of HEP, for example in New Zealand a recent drought has led to a supply shortage.

e **10/10 marks awarded (grade A).** This answer is well balanced across the six renewable sources, with a good grasp of physical factors and good use of examples. It shows sound, up-to-date knowledge and uses clear explanations. It is also well structured, with good use of geographical terminology. The answer makes good use of the whole of Figure 1. **a** A good explanation of biofuels, showing knowledge and understanding of a range of sources, but not fully linked to the size of biofuel potential shown in the Figure. **b** This section on solar power is linked to the 'largest' potential on the Figure. **c** Some good details on wind, which show an understanding of how its widespread use and development can explain the large wind component on Figure 1. **d** This section might have been more logically placed with the first paragraph on biofuels. **e** The limitations of energy sources with little potential are explained, with good reference to examples.

(b) Assess the role that fossil fuels might play in future energy security. (15 marks)

e Consider coal, oil and gas and the extent to which developing new and existing resources might improve energy security in named countries.

Mark scheme and levels

e Fossil fuels are likely to be the main source of energy for the foreseeable future. In some countries with large reserves (Chinese coal, Russian gas, Middle East oil) they will play the major role. In the UK dependency on gas is set to grow, despite declining domestic supplies. Gas is seen as clean compared with coal, and helps the UK meet its Kyoto and air quality targets; gas is increasingly being imported as LNG from Qatar, which has huge reserves. In countries without domestic supplies there may be increasing moves towards renewable sources; this has already happened in France with nuclear power and is happening with biofuels in the USA. Technology has a role to play, as it is not certain that viable electric or hydrogen vehicles can be developed. Countries need to have alternatives and this depends on the availability of alternative sources; the UK has large wind and tidal potential but other countries may have fewer options.

Level	Mark	Descriptor
Level 4	13–15	Detailed, structured answer that uses a range of examples and real situations. Good understanding of energy security and energy sources; genuine assessment. Explanations are always clear. Geographical terminology is used with accuracy. Grammar, punctuation and spelling errors are very rare.
Level 3	9–12	Some detail in a structured account that uses some examples to illustrate some aspects of energy security. Explanations are always clear. Geographical terminology is used with accuracy. Grammar, punctuation and spelling errors are rare.
Level 2	5–8	Limited structure and detail; unbalanced towards particular factors or examples. Lacking in range. Explanations are clear, but there are areas of less clarity. Geographical terminology is used with some accuracy. There are some grammar, punctuation and spelling errors.
Level 1	1–4	One or two generalised points perhaps linked to one location. Explanations are oversimplified and lack clarity. Geographical terminology is rarely used with accuracy. There are frequent grammar, punctuation and spelling errors.

Student answer

There are three major fossil fuels, namely oil, gas and coal. Currently fossil fuels contribute 85% of global energy supply (coal 25%, oil 37% and gas 23% respectively) but there are a number of issues that mean that fossil fuels may have a diminished role in future energy security. **a** In the first place fossil fuels, especially coal and oil, but to a lesser extent gas, make a major contribution to greenhouse gases and the problem of climate change so there is environmental controversy about their future. **b**

A second consideration is that fossil fuels, especially oil, are declining in supply — many people argue that we have actually reached peak oil, and that peak gas is perhaps only 50 years away. Coal reserves will probably last for around 200 years at current use rates, but coal is less energy dense than oil or gas, more costly to transport and dirtier — especially in terms of acid rain caused by sulphur dioxide. **c**

A third consideration is that the geographical occurrence of both oil and gas, even with new discoveries, and the use of more unconventional oil supplies such has the Athabasca tar sands, lead to remarkable concentration, especially in the Middle East and Russia (gas). Therefore geopolitical considerations impact on the security of supply — hence the concern when Russia cut off gas supplies to Ukraine in 2008, and Belarus in 2009, having an impact on security of supply for countries such as Slovakia in Eastern Europe. Moreover many of the pathways along which fossil fuels are moved are currently insecure. **d**

A fourth consideration is that even with a global recession in 2008–2010 economic growth especially in China and India is likely to lead to rising demand for fossil fuels — China in particular is contributing to the development of new oil supplies in both Latin America and Africa in order to secure supplies. Demand has a major impact on the price of a barrel of oil — high prices for oil mean that alternative sources, including unconventional oil, nuclear power and renewables, all become much more economically viable to develop. **e**

A fifth consideration is the role of technology — carbon capture could revolutionise how coal is viewed environmentally, although there will still be issues of mining costs and environmental damage. Technology could also lead to the development of alternatives to the internal combustion engine.

The statement cannot only be considered globally, but also with reference to a number of different groups of countries. **f** If we look at the BRICS for example, it is highly likely that Russia, China and India will continue to rely substantially on fossil fuels for the foreseeable future to overcome energy security issues. Brazil is unusual in that in response to a lack of major supplies of fossil fuels it has become a lead developer of biofuel from soya beans and sugar cane, with disastrous impacts on forest biodiversity. It would be equally true to say that OPEC countries will continue to rely on their oil and gas supplies for the foreseeable future as they have high security. **g**

However, energy security has many facets and also can change over time. While the USA has many supplies of indigenous fossil fuels, for geopolitical reasons it is increasingly developing new supplies in environmentally sensitive locations such as the ANWR and north Alaska, and the recent environmental disaster in the Gulf of Mexico has put into question drilling for oil in deep under-sea locations. For this reason the USA has made a massive move towards biofuels. Recently the whole picture of gas security in the USA has completely changed with the discovery of vast supplies of shale gas in states such as Pennsylvania.

Other countries with lower energy security, such as Britain post North Sea, or France, are looking to diversify their energy mix considerably and for these countries it may well be that fossil fuels will play a much diminished role in future energy security. France for instance has invested heavily in nuclear power, whereas in the UK there has been a huge investment in wind power, and also moves have been made to secure LPG supplies from 'friendly' states such as Qatar. **h**

In conclusion, therefore, globally it would be possible to state that fossil fuels will play a very important part in future energy security for the foreseeable future, but that in certain countries, for a variety of reasons there is a move to develop energy strategies that are less fossil fuel dependent. **i**

ⓔ **15/15 marks awarded (grade A).** This is a wide-ranging answer which does try to assess the statement. It is well supported by current examples and shows a contemporary understanding of energy issues. It is well structured with good use of geographical terminology. **a** A good introduction, which uses some hard facts and defines fossil fuels. **b** A key point about fossil fuel use is made here. **c** A detailed discussion of 'peak issues'. **d** This section has some good examples of where supplies are located and supply pathways. **e** Some details of the economics of fossil fuels are useful. **f** The answer provides some assessment here, by recognising that it is not just the global picture that has to be examined. **g** This section on different countries begins to show that the question of the role of fossil fuels is actually a complex one. **h** Throughout, this answer uses a range of examples but avoids being side-tracked into major case studies. **i** A good overall conclusion, which assesses the global picture against the situation in individual nations.

Question 2 **Water conflicts**

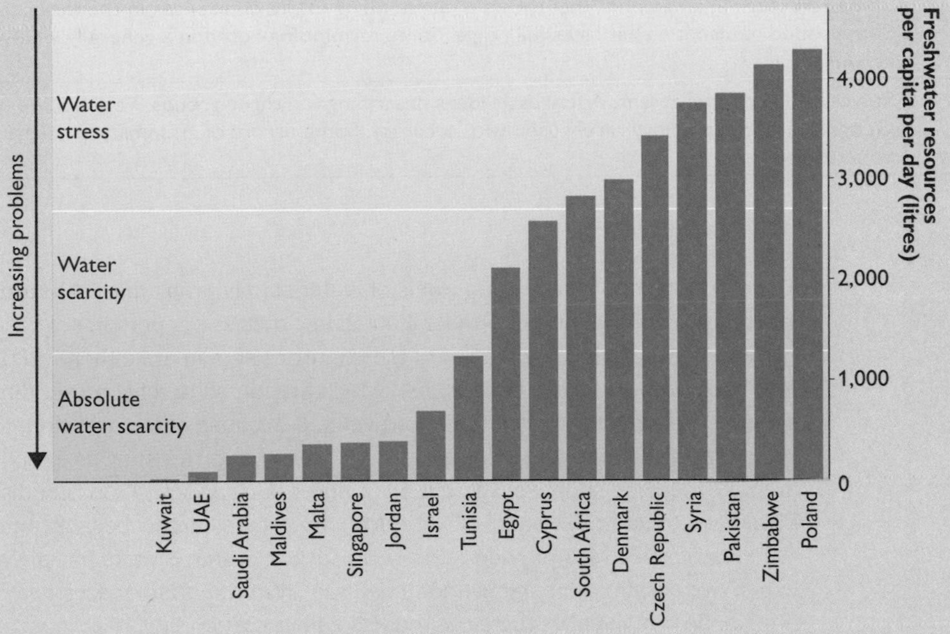

Figure 2 Freshwater resources

(a) Using your own knowledge and Figure 2, explain why the countries shown experience water supply problems. (10 marks)

ⓔ Focus on a range of countries in Figure 2, across the graph. Your reasons should include physical factors and economic issues. Try to use 'water' terminology.

Mark scheme and levels

ⓔ Define terms (use definitions from resource). Note: in litres per day, not cubic metres per year as is usual. Group countries:

- absolute scarcity — 6/9 are in the arid/semi-arid Middle East/North Africa; 3/9 are island nations: Malta (limestone), Maldives (salt incursions/coral atolls) and Singapore (crowded island state)
- scarcity — Egypt (river Nile — still desert), Cyprus (Mediterranean)
- water stress — wide range of countries, some semi-arid, e.g. South Africa, Syria (physical factors); economic factors could be an issue (Zimbabwe — breakdown of economy and infrastructure)

Level	Mark	Descriptor
Level 3	8–10	Well structured. Clear understanding of three categories with sound reasons for a range of countries. Good use of terminology. Well written.
Level 2	5–7	Some structure in an account that shows understanding of the differences in the three categories. Some sound explanation but lacks full range. Some terminology used in a generally well-written account.
Level 1	1–4	Structure is poor or absent. A few basic ideas describing the three groups. Very limited explanation. Geographical terminology rarely used with accuracy. Some errors of grammar, spelling and punctuation.

Student answer

The diagram shows increasing levels of water supply problems. **a** **Absolute scarcity**, usually defined as under 1,000 cubic metres per person per year is an extreme state of deficiency. Most of the countries shown are either arid or semi-arid such as those in the Middle East, which are forced to rely on ever diminishing supplies from rivers, lakes and groundwater. **b** Because of the arid climate, and growing populations and irrigated agriculture in countries such as Israel, demands are far exceeding supplies. The other countries such as Malta are island states and have been traditionally short of water for many years, largely because of the geology — limestone — which provides almost no surface water. **c** Malta has always relied on expensive desalination plants and the all year round tourism places heavy demands on supplies. In all cases therefore there is a physical scarcity.

Scarcity **a** is a slightly less dire situation. Some countries such as Egypt, where 92% of people live in the Nile Valley, are clearly dependent on the health of the only significant perennial river. The Aswan Dam has regulated supplies, providing water for irrigation of food crops to feed the growing population. However, further development of upstream dams in the Ethiopian highlands may lead to a dropping of supplies in the Blue Nile. **d**

Water stress **a** (under 1,700 cubic metres per person per year) is a category which many countries will be joining largely because of the impacts of climate change **e** which will certainly lead to more uncertain weather and increasing droughts, for example in South Africa, making it extremely vulnerable as it has a growing increasingly unbalanced population and is also dependent on irrigated cash crops for export. Some of the listed countries such as Zimbabwe are almost certainly experiencing economic water scarcity as a result of the economic breakdown and increasing numbers of people living in poverty who are unable to afford the rising cost of privatised water supplies. **f**

e **9/10 marks awarded (grade A).** Overall, this is a good answer as it is well structured and uses terminology effectively, with exemplar support. **a** The student shows good understanding of the three categories. This provides a sound structure for the answer. **b/c** The answer provides accurate and relative details of physical reasons for absolute scarcity, although there is no mention of the Maldives. **d** Useful one-country information is supplied on Egypt for scarcity. **e/f** Vulnerability to climate change and to economic breakdown are both well developed.

(b) Assess the extent to which plentiful supplies of water in some parts of the world can be used to make up shortages elsewhere. (15 marks)

e 'Assess' means weigh-up, so you should consider the pros and cons of transferring water from areas of surplus to areas of shortage; try to use national and transboundary examples.

Mark scheme and levels

e Traditionally water has often been moved within a country — water grid, or building reservoirs in hilly areas then pipelines to urban areas, e.g. Colorado reservoirs to Los Angeles, California. Water tankers go round the world serving small island communities, e.g. Bahamas. Sometimes polyurethane bags (up to 2 million litres) are towed out to Greek islands. Currently water transfers are the main mechanism — politically straightforward within a country, but environmentally difficult (e.g. Tagus). Issues for source/receiver. More difficult internationally. Name current examples of international transfer (e.g. Israel). Proposed transfers are often very large scale, e.g. South–North in China. Are they environmentally safe and economically viable? As the question requires assessment, a good answer will include details of problem areas too. Many areas' physical/economic circumstances make transfer non-viable or not feasible. So conservation strategies could be more useful. Also with climate change areas of plenty are declining. Issues of virtual water. Many areas of plenty, e.g. UK, actually import food from irrigated areas elsewhere — so they have a very large water footprint.

Level	Mark	Descriptor
Level 4	13–15	A well-structured assessment of the issues of water transfer, supported by a range of detailed examples. Sound use of terminology. Well written, with good standards of spelling, punctuation and grammar.
Level 3	9–12	Begins to assess in a structured account that shows how plentiful water can and cannot be used to make up shortages elsewhere. Relevant exemplification, e.g. of water transfer. Some use of terminology. Generally well written.
Level 2	5–8	Some structure in a descriptive account that describes how water can be transferred from areas of plenty to areas of scarcity. Does give examples but lacks range. Quite clearly written, some use of terminology but some technical errors.
Level 1	1–4	One or two basic statements in an account that lacks structure. Examples used are not well located and not always relevant. Lacks terminology. There are frequent grammar, spelling and punctuation errors.

Student answer

Some areas of the world do not suffer from water shortages at all and can be said to have surplus water. For water stressed regions, this surplus is an obvious solution but transferring water is rarely sustainable politically, environmentally and sometimes not economically. **a** When water resources are transboundary, either rivers or aquifers, **b** there are often political problems with making transfers. This is the case with Israel, which is accused of overusing water from the River Jordan and draining the West Bank mountain aquifer which lies under Palestine. Israel takes 80% of its water and the conflict over this contributes to the political instability between the two countries. **c** Sometimes transboundary resources can be used and shared but only when all players sign up to an agreement such as the Helsinki Rules — as is the case with the Mekong River Commission. **d** In theory transferring water within a country should be easier but as the Colorado River shows this is not always the case. So much water has been extracted from the Colorado that its discharge is basically zero when it enters the sea, and the river ecosystem has been devastated. A similar situation exists in the Aral Sea where Soviet engineers diverted so much water that the sea shrank catastrophically, devastating the fishing industry as the lake's ecosystem collapsed. These examples show that water transfers usually lead to ecological disaster and are not environmentally sustainable. **e**

China is currently implementing a vast engineering scheme called the South–North Water Transfer. The Three Gorges Dam is one part of this and there will also be other dams, canals and river diversions taking water from the water surplus in the south towards the dry areas around Beijing in the north. The scheme is planned to take 50 years to complete and cost $billions. **f** There are question marks over whether this is a good way to spend money and whether China can afford it. Often this type of scheme eventually turns out to be a poor way of securing a water supply. In the USA there are fears that the falling level of Lake Mead behind the Hoover Dam will cut off the supply of water to Las Vegas. Global warming and changing rainfall patterns have probably contributed to this situation but were not factors when the water transfers was originally built — the same thing could occur in China. **g**

Overall, areas which do not have plentiful water supplies should not try to take surplus water from elsewhere. As the examples used have shown, transfers lead to political conflict, environmental degradation and might turn out to be expensive 'white elephants' due to climate change. Water stressed regions should focus on water conservation and using the supply they have in a more sustainable way. **h**

e 13/15 marks awarded (grade A). This is a good answer. It has some range of examples and some details, although it is a little thin on hard facts. It is well structured. **a** A good introduction, which sets out a structure for the rest of the answer. **b** Good use of terminology. **c** This is a good example, although it could be more detailed. **d** A good assessment here, although more depth on the Mekong situation would help. **e** There are two useful examples used in this section, plus an overall assessment of the environmental consequences of transfers. **f** There is some factual detail here, but it could be more accurate. **g** This is a good point, using two examples to argue that the high economic cost of transfers might be undone by climate change in the future. **h** This is a very good summary, which comes to a firm conclusion in direct response to the question.

Question 3 **Biodiversity under threat**

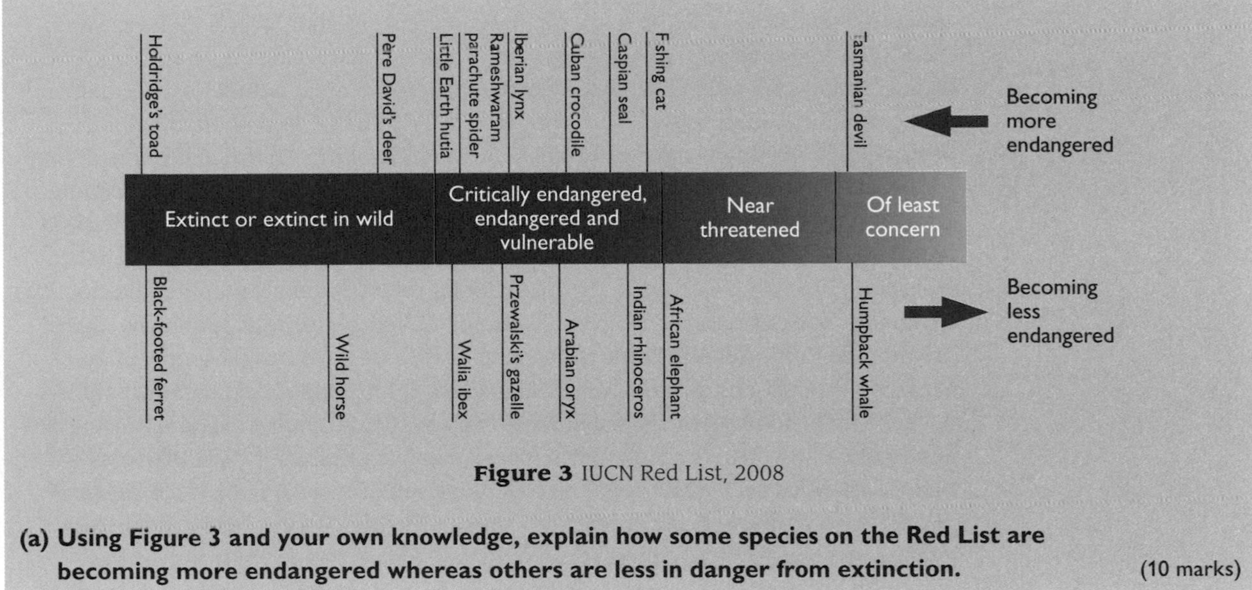

Figure 3 IUCN Red List, 2008

(a) Using Figure 3 and your own knowledge, explain how some species on the Red List are becoming more endangered whereas others are less in danger from extinction. (10 marks)

ⓔ You should make frequent reference to Figure 3, and try to use some of your own ideas/examples. Try to include a range of reasons — perhaps three.

Mark scheme and levels

ⓔ IUCN Red List refers largely to mammals (easy to count). Species are becoming endangered and less endangered in all categories of the Red List (give examples). Greater numbers are in the extinction and critically endangered category than in the least concern category. Explanations will therefore be linked to examples of damaging actions, and also to successful conservation strategies (no need to refer to species).

Level	Mark	Descriptor
Level 3	8–10	Well-balanced, structured account, with detailed exemplification. Explanations are always clear. Geographical terminology is used with accuracy. Grammar, punctuation and spelling errors are rare.
Level 2	5–7	Some structure in an account with some understanding. Explanations are usually clear. Lacks full range and may be unbalanced. Some exemplification. There are some grammar, spelling and punctuation errors.
Level 1	1–4	Lacks structure. Limited understanding of topic. Explanations are simple and basic and lack clarity. Limited geographical terminology. Frequent grammar, spelling and punctuation errors.

Student answer

The red list is one way of seeing how threatened biodiversity is as it is done annually placing species, usually mammals, **a** in one of ten categories from extinct to little concern. Figure 3 shows that it is a rapidly changing situation **b** as at all levels on the scale some species are becoming more endangered **c** and some less so, such as the humpback whale campaign (Greenpeace). There are a number of reasons why species are becoming more endangered. **d** It can be to do with their habitat becoming more endangered, for example the melting of the Arctic ice has meant that polar bears are under threat as they have less opportunities for feeding or breeding. Some animals because of their large size are very vulnerable to hunters or they may have very valuable resources such as ivory from elephant's tusks, and only strict legislation like CITES can make them less under threat. Animals that are unattractive to humans such as rats and mice are in danger of being wiped out. Equally badgers are thought to cause TB in cattle, so in spite of various Save Badger groups they are trapped or shot by farmers. Yet other species are becoming endangered because of the impacts of alien species — such as the red squirrel, which is being systematically overrun by grey squirrels and is only found in northern Britain now. Endemic species such as those in the Galapagos are especially vulnerable because of their comparatively small populations and inbreeding. Species can become less endangered, for example if there are campaigns to save them. Often as was the case with the giant panda, which was brought back from the brink, conservation strategies have to be combined with captive breeding in zoos. **e**

e 8/10 marks awarded (grade A). Overall there is some structure, some use of terminology (endemics and captive breeding) and satisfactory standards of quality of written communication. **a** The student understands what the Red List is. **b/c** The student understands the implications of Figure 3 and gives one example drawn from it. **d** The answer establishes a range of exemplified reasons as to why species are becoming more endangered. **e** Unfortunately the student only gives one example as to why species are becoming less endangered, so the answer is a little unbalanced.

(b) Evaluate the link between economic development and ecosystem destruction. **(15 marks)**

e This question will benefit from a clear definition of economic development and ecosystem destruction. Use named examples and focus on countries at different levels of economic development, and explore whether as development increases more destruction is inevitable, or not.

Mark scheme and levels

e Economic development leads to the development of resources, energy and commercial farming, which can all have impacts on ecosystem destruction. As these resources are processed into manufactured articles, there are many indirect impacts such as air and water pollution (e.g. rising GHG, acid rain, destruction of ozone layer). Further development leads to reliance on tertiary/quaternary industries, and increasing reliance on transport and energy as people become wealthier and consume more. In theory, greater wealth should lead to more awareness of sustainable solutions (money and technology) within the country, but in some cases the biodiversity loss is exported to developing countries.

Level	Mark	Descriptor
Level 4	13–15	Well-structured evaluation of the links, well supported by examples. Explanations are clear, with good use of geographical terminology. Grammar, spelling and punctuation errors are rare.
Level 3	9–12	Structured account with limited evaluation, supported by examples. Explanations are clear, with accurate use of terminology. Grammar, punctuation and spelling errors are rare.
Level 2	5–8	Some structure in an account that gives examples of the links. Explanations are usually clear. There are some grammar, punctuation and spelling errors.
Level 1	1–4	Lacks structure; simple basic explanation. Rare use of geographical terminology. Frequent grammar, punctuation and spelling errors.

Student answer

The relationship between economic development and ecosystem destruction can be summed up by the following diagram. **a**

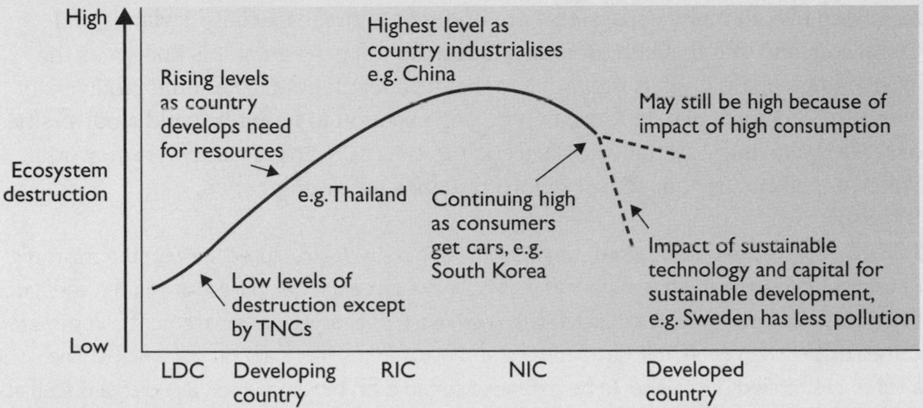

If you start with least developed countries such as the Cameroon Republic they have very low levels of industrialisation, but need to export forest and mineral resources, to earn income. This leads to rainforest destruction, e.g. in Korup. As the population is rising in LDCs this exerts pressure on the land, as people chop down rainforest to subsist, or use coral reefs for fishing as in the Philippines. **b**

As countries develop, and begin to industrialise, this leads to rising demands for resources, and also a need for energy. In China they build one coal-fired power station each week, and levels of both air pollution and water pollution are exceedingly high, again damaging ecosystems by acid rain (Yunnan forests). While there are attempts to conserve species such as the panda, the pace of development is so great that ecosystems are being destroyed at an amazing rate. As these countries become more affluent their people have more purchasing power and they buy motor cars, so adding to greenhouse gas pollution — this phenomenon is global as it is happening all round the world. The resultant climate change is considered to be the greatest threat of all to biodiversity. **c**

When countries become more developed, theoretically they have the capital and technology to develop sustainable management and to conserve large areas of their valuable ecosystems. For some systems such as temperate forests and grasslands it is really too late, as up to 70% have been destroyed. Many countries however carry on as usual and ecosystem destruction (especially in lakes and rivers) continues as a result of eutrophication and other pollution. Some developing countries such as Tanzania and Costa Rica value their ecosystems and have excellent conservation strategies. **d**

Overall the pattern of ecosystem destruction can be matched to global economic development as, because of globalisation, systems are interconnected. Many of the ecosystems in tropical areas are under greatest threat, for example rainforest and reefs. Rainforests are being threatened for timber, and also are being chopped down and replaced by soya bean plantations (Brazil), or palm oil plantations (Indonesia) for producing biofuels. Globalisation has also led to world-wide tourism, and also industrial-level fishing, both of which have led to increased pressure on reefs which are already, for example in St Lucia, experiencing damage from development-related pollution and siltation. **e**

In conclusion, therefore, there is undoubtedly a strong link between economic development and ecosystem destruction. The Millennium Ecosystems Assessment suggested that all major ecosystems are under pressure, especially drylands and freshwater, and that this human-induced destruction is so great it is known as the sixth extinction. So great is the problem that as a result of International Biodiversity Year 2010, there is a plan to form an IPCC organisation to monitor world biodiversity loss. There are many systems in place such as the IUCN Conservation Frameworks, which should ensure that conservation strategies are strengthened.

e 13/15 marks awarded (grade A). Overall this is a well-structured answer; the diagram is very useful in providing this framework. It is also an evaluative answer, particularly in the very good conclusion. The answer shows an up-to-date knowledge of ecosystem management and shows that the link between economic development and ecosystems is actually quite complex. The answer is exemplified but this could be extended for an even better answer. **a** A diagram such as this is a very useful way of summarising an argument. **b** Some good examples are used here and these link to the summary diagram that the student has drawn. **c** This section has moved on to the rampant ecosystem destruction often seen in NICs. **d** There is some evaluation here, because it is recognised that some developing countries conserve, whereas some developed countries continue to degrade. **e** This section is good, because it makes the link between globalisation of economic activity and globalised destruction of ecosystems.

Question 4 **Superpower geographies**

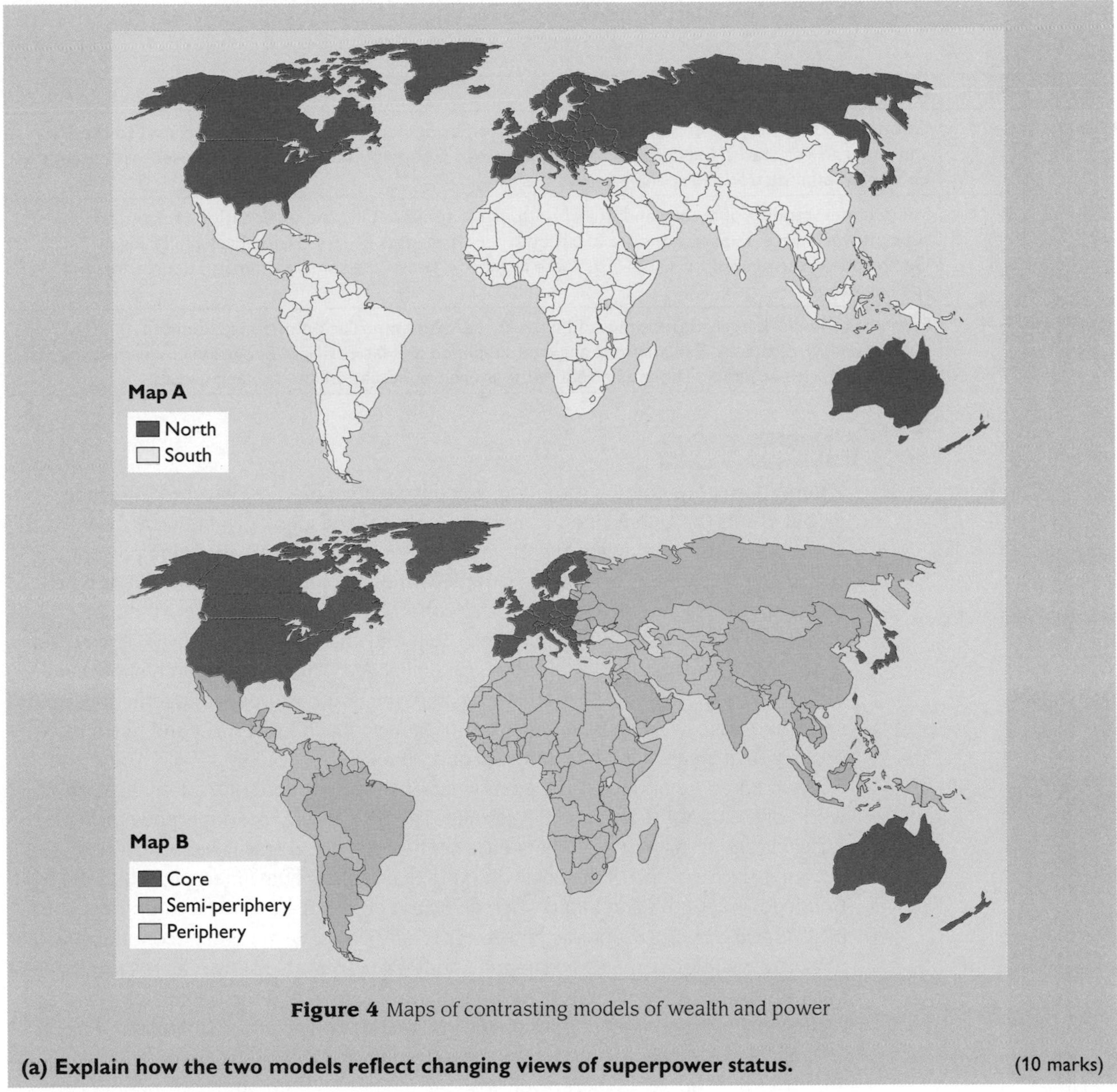

Figure 4 Maps of contrasting models of wealth and power

(a) Explain how the two models reflect changing views of superpower status. (10 marks)

ⓔ It is important to have balance between the two models, and focus on how they might help explain patterns of superpowers now and in the past and/or future.

Mark scheme and levels

ⓔ The North–South divide dates from the 1970s/1980s and represents a polarised view of the world. A North–South world has power concentrated in North America, Europe and Russia — it is a Cold War model with the USA dominating the West and USSR the East, and both dominating the South. It views the entire South as basically the same. Wallerstein's model sees the world as falling into three divisions of core, semi-periphery and periphery. World systems theory includes NICs as countries that are growing, but still tied to the core. It is possible to see the BRICs in Map B, but not in Map A. Map B might show a world closer to reality in 2011 than Map A, as the emerging superpowers can be identified in Map B but not on Map A.

Level	Mark	Descriptor
Level 3	8–10	Good understanding and comments with balance on both maps. Explanations are related to theory and superpower status. Explanations are always clear. Geographical terminology is used with accuracy. Grammar, punctuation and spelling errors are rare.
Level 2	5–7	Some understanding of theory and use of both maps; some details and explanations related to superpower status. Explanations are clear, but there are areas of less clarity. Lacks full range. Geographical terminology is used with some accuracy. There are some grammar, punctuation and spelling errors.
Level 1	1–4	Some basic ideas. Largely descriptive and unbalanced with emphasis on one map. Limited understanding of theory. Explanations are oversimplified and lack clarity. Geographical terminology is rarely used with accuracy. There are frequent grammar, punctuation and spelling errors.

Student answer

A superpower is a country, or group of countries, that has the power to influence global events through economic, military, cultural and ideological power. Superpower status changes over time as superpowers fall and emerging powers rise. **a** Map A is an out-of-date view of the world. It could date from the 1980s when the idea of North versus South was very common. In the North there are the USA and NATO countries of Europe plus the USSR and eastern bloc countries. This world order ended in about 1990 when the USSR collapsed. This means Map A shows an out-of-date geography. **b** Also on Map A all of the South is grouped together but in fact some places such as sub-Saharan Africa have much less power and much more poverty than areas like South America and China. **c**

Map B is more about today's geopolitics. It shows the core of the rich western world of the USA, EU, Japan and Australia. The USA and EU are superpowers, but Russia is shown as part of Wallerstein's semi-periphery. **d** This is because it has declined since the end of the cold war. Map B also shows the emerging BRICS of Brazil, Russia, China and India, whereas Map A does not show these countries. The BRICS have growing economic power, especially China. Some other economically powerful counties such as Mexico and Saudi Arabia are also shown as NICs in the semi-periphery. **e** The periphery includes the poorest area in the world (sub-Saharan Africa) and also some other poor countries such as Bolivia and Afghanistan. **f** Overall, Map B shows the modern world dominated by the western capitalist countries and the USA superpower, with the emerging BRICS and poor periphery. Map A is an out-of-date map. **g**

e 10/10 marks awarded (grade A). This is a very good answer. It is structured, with a brief introduction that defines what a superpower is, balanced sections on Map A then Map B, followed by a brief summary. Good geographical terminology is used throughout. There are links to theory and the geography of the whole of the two maps in covered with detailed reference to countries. **a** A good definition of superpowers, which demonstrates understanding. There is also a recognition that status changes, which is what the question is focused on. **b** This section is very good, as the student makes a strong case for Map A being a map about a past geography. **c** This is a good point, as it shows the student is considering the whole of map A. **d** Good link to Wallerstein's world systems theory. **e** There is a good discussion of the BRICs linked to the theory shown in Figure 4. **f** This part discusses the periphery, so all parts of Figure 4 have been explained. **g** A useful summary at the end.

(b) Assess the role of trade and international organisations in maintaining superpower status. (15 marks)

e Focus on how important trade and international decision making are (or are not) in terms of helping a superpower maintain or gain status; you could mention other factors (culture, military power) and discuss how important they are.

Mark scheme and levels

e Trade is important as it is often called the 'engine of growth'. It creates the wealth superpowers need to support their global networks. There are arguments that terms of trade keep the rich powerful and the poor weak (the neo-colonial/dependency model). China could be seen as an example of a country that has used trade to increase wealth and enhance its status as an emerging power. Superpowers dominate international organisations and influence their decision making. International organisations can be linked to trade as trade blocs such as NAFTA and the EU work in the interests of free trade between rich countries, but might be seen to exclude the less powerful from these benefits by imposing tariffs and quotas. Membership of the UN security council, WTO, NATO and G8 give the superpowers an important lever in global decision making and help them 'get their way'. The emerging powers/BRICs would see their power and influence enhanced if they became members of some IGOs. Other factors such as military power and cultural influence are also important in maintaining power and the answer could be broadened to discuss these issues.

Level	Mark	Descriptor
Level 4	13–15	Balanced across trade and international organisations. Good real world details and examples for both; other factors mentioned in a genuine assessment. Explanations are always clear. Geographical terminology is used with accuracy. Grammar, punctuation and spelling errors are very rare.
Level 3	9–12	Some balance across trade and organisations, and some details for both. At lower end of level, emphasis on only one. Some explanations and implied assessment. Explanations are always clear. Geographical terminology is used with accuracy. Grammar, punctuation and spelling errors are rare.
Level 2	5–8	Unbalanced; some explanations stronger than others and there is some detail but patchy. Explanations are clear, but there are areas of less clarity. Geographical terminology is used with some accuracy. There are some grammar, punctuation and spelling errors.
Level 1	1–4	A few general ideas, most likely on trade and lacking supporting evidence. Explanations are oversimplified and lack clarity. Geographical terminology is rarely used with accuracy. There are frequent grammar, punctuation and spelling errors.

Student answer

Superpowers maintain their status in a number of ways. These include military strength, cultural, geopolitical and ideological influence and economic power. Economic power is important because without it superpowers could not afford military power and would find acting on the world stage difficult. **a**

Trade is one of the key ways in which superpowers make money. Most world trade takes places between Europe, North America and Asia. This is trade in goods and services and is high value. It earns money for global TNCs, the majority of which are from the developed world. Out of the 500 largest TNCs in 2010, 140 were from the USA and they had combined revenues of nearly $7 trillion. **b** This huge wealth allows the USA to afford the most technologically advanced, and most global, military in the world. For instance the USA has 11 nuclear-powered aircraft carriers. **c**

Much of the trade the USA carries out is protected by the patent system. This means that when TNCs invest in research and development and bring out new products, only those TNCs benefit because copying their ideas is illegal. **d** TNCs also mean that the USA spreads its cultural influence around the world through globally known brands like Coca-Cola, CNN and Disney. These brands sell the idea of the 'American dream'. **e** In addition, the USA is an influential member of the World Trade Organization. This inter-governmental organisation has promoted free-trade. This has led to an increase in world trade which has benefited the EU and USA. More world trade basically means greater profits for TNCs who can take advantage of trade. The USA is the key player in the NAFTA trade bloc, which has free trade agreements with the EU. Trade blocs make trade easier and more profitable and the USA has gained economic power from this. **f**

China has benefited from world trade as its economy has grown by 8–12% per year since 1990. This has meant huge increases in wealth in China but it is not yet a superpower. Wealth alone is not enough as influence is needed and China lacks this compared with the USA. **g** Besides the WTO, other international organisations are important for superpower status. The USA is the biggest contributor to the UN budget and has a seat on the UN Security Council. This means it can influence global political situations such as applying sanctions to Iran or influencing the Israel–Palestine conflict. The USA influences global economic policy as it has 17% of the votes at the IMF. This gives it the power to project its free-market ideas around the world, for instance forcing developing countries that are in debt to make free market reforms before getting debt relief **h**.

Overall, trade creates money and money is needed to be a superpower. Wealth means that a country like the USA can buy military hardware and make itself powerful. International organisations are important for influencing geopolitical decisions, and making alliances. Trade and IGOs are not the whole story because cultural influence is important too. This is what China lacks at this time despite its growing wealth. **i**

e 15/15 marks awarded (grade A). This is a very good answer, packed with examples and understanding. It has a balance between trade and international organisations, as well as bringing in other factors such as culture and military strength. It is a real assessment, as it weighs up the importance of trade and international organisations and has a clear conclusion. **a** Introduction focuses on superpower status, and makes the case for the importance of economic power. **b** Good focus on trade, and factually very detailed; a few accurate facts and figures are a good way to impress. **c** A good link between trade, wealth and military might; shows understanding. **d** The point about the patent system is well made. **e** The discussion of TNCs and trade is expanded into cultural influence. **f** This section is very good. It links together trade and international organisations using the WTO as an example, and then links to named trade blocs. **g** China is used as an example to argue that trade alone is not enough to make a superpower, which is the assessment the question calls for. **h** Good examples of IGOs and the influence they bring. **i** A good concluding statement, which assesses the value of trade against other influences on superpower status such as IGOs and culture.

Question 5 **Bridging the development gap**

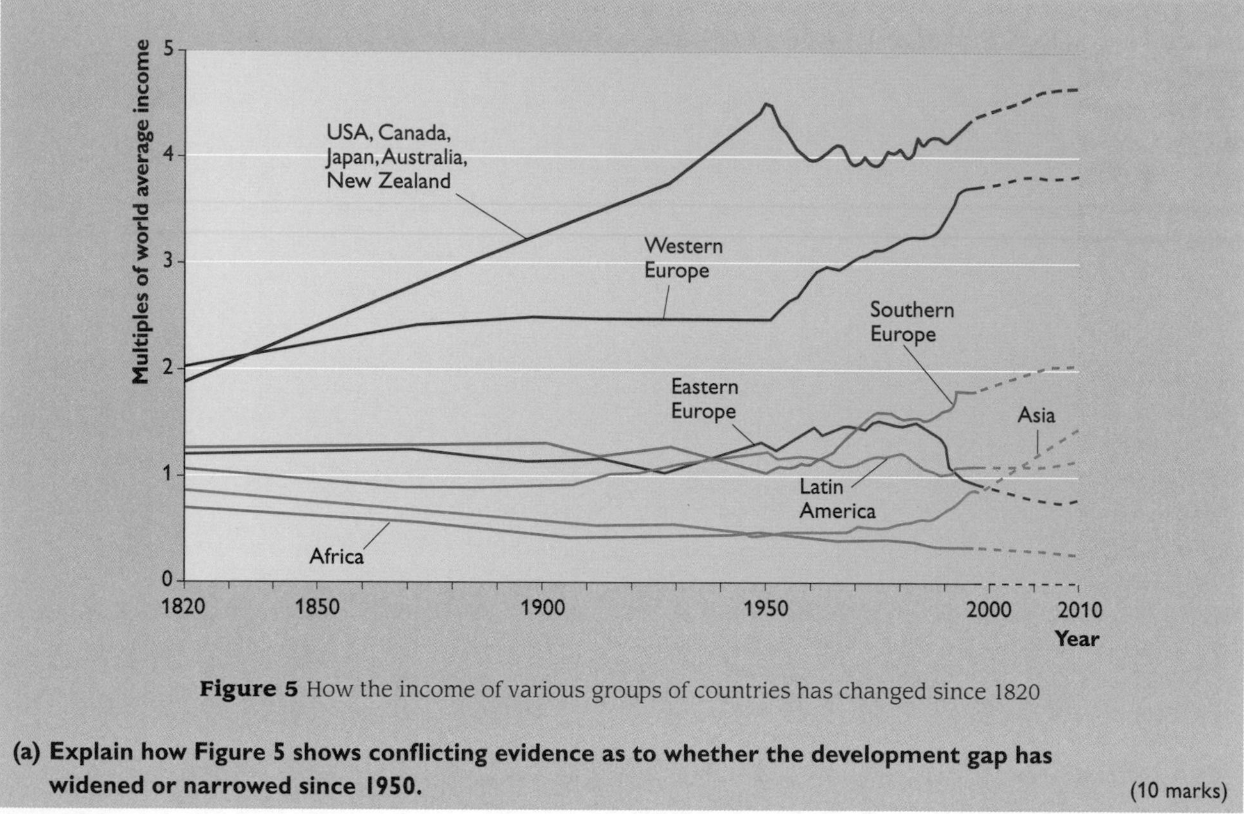

Figure 5 How the income of various groups of countries has changed since 1820

(a) Explain how Figure 5 shows conflicting evidence as to whether the development gap has widened or narrowed since 1950. (10 marks)

e The key word here is 'conflicting'; you should explain how Figure 5 suggests both narrowing and widening and have some balance between these. Refer to the data directly.

Mark scheme and levels

e Since 1960 a number of groups have raised their incomes by 1–1.5 times the average world income, notably west and south Europe, Asia and non-European MEDCs. Reasons could include continued investment, trade etc. Other areas such as Latin America have remained static — a mix of economies from Brazil (wealthiest) to others that are very poor. Africa — mainly a continent of LDCs — is a very poor continent and its steady decline suggests a widening of the development gap. Eastern Europe also shows a marked decline (upheaval due to the break-up of the communist states) and indicates some widening of the development gap with reference to MEDCs. The spectacular climb of Asia — NICs' rapid industrialisation — suggests a narrowing of the development gap.

Level	Mark	Descriptor
Level 3	8–10	Provides a detailed, balanced, structured account across Figure 5. Explanations are clear. Geographical terminology is well used. Grammar, punctuation and spelling errors are rare.
Level 2	5–7	Some structure in an account that looks at both sides but needs more detail. Explanations are usually clear. Some geographical terminology used. Grammar, punctuation and spelling are satisfactory.
Level 1	1–4	Lacks structure, which shows limited understanding of resource. Basic, simple explanations. Frequent grammar, punctuation and spelling errors.

Student answer

The development gap is the gap between the richest and poorest. It is often viewed as the gap between the developed and developing world. Figure 5 has evidence that the gap is widening and narrowing at the same time. **a**

Since 1950, the developed world has pulled away from other parts of the world. Western Europe's multiple of world average income has risen from 2.5 in 1950 to close to 4 in 2010. Other developed countries (Canada, Japan, USA) have also increased from 4 to 4.75. Overall, these developed countries have pulled away from the developing countries beneath them, widening the gap. **b**

The most obvious sign of a widening gap is the fact that Africa's level has dropped slowly but continuously since 1950, suggesting it has become poorer in relation to the rest of the world. This is despite the fact that it was the poorest region in 1950. It has made no progress at all. Eastern Europe seems to have become poorer probably as a result of the collapse of communism in 1990. Recently it has picked up a little since some countries joined the EU. **c**

However, there is strong evidence that the gap is narrowing in some regions. Asia saw a huge rise in incomes from 0.5 in 1950 to 1.5 in 2010 suggesting countries such as China and India are getting wealthier. Despite this, Asia has not really closed the gap on Western Europe, it has just kept pace. It has overtaken Latin America, which seems to have stagnated. **d**

Overall, Europe (western and southern), Asia and other OECD countries seem to have become richer, but Eastern Europe, Latin America and Africa seem to have declined or stagnated, leading to a complicated picture of narrowing and widening. **e**

e **10/10 marks awarded (grade A).** This is a very good answer which uses the Figure with precision. It refers to all regions, uses data and makes specific reference to narrowing and widening. **a** Good definition of the development gap with reference to Figure 5. **b** This section uses data from Figure 5 to support the idea of widening. **c** This section supports the idea of widening, and shows good real world understanding of Eastern Europe. **d** This section has a correct and complex argument showing that Figure 5 is being used carefully, and fully. **e** A useful overall summary.

(b) Evaluate the role of aid in bridging the development gap. (15 marks)

e Your answer needs to use examples of aid, ideally a range of types of aid. Consider the extent to which the examples have helped the development process for different groups of people. Named examples are not specifically asked for, but you do need to use them.

Mark scheme and levels

e Definitions of aid, looking at the spectrum, which includes humanitarian aid in emergencies, debt relief, technical assistance etc. Classifying aid into bilateral, multilateral, tied aid etc. and making the point that the type of aid may determine how successful it is in bridging the development gap. Top-down versus bottom-up is relevant here. Support for aid: saving lives in emergencies, acting as a pump primer for development, redistribution of global wealth, establishing practical links between countries (interdependence). Opinions against aid: can be misused by corrupt governments to favour the rulers, can create dependency, can distort free market, can promote attitudes of superiority, can be used for strategic, political and economic reasons — tends to maintain world inequality, not challenge it.

Level	Mark	Descriptor
Level 4	13–15	Well-structured account, which evaluates the role of aid with detailed exemplification. Explanations are always clear with good use of geographical terminology. Very good standards of quality of written communication.
Level 3	9–12	Structured account that begins to evaluate the role of aid. Explanations are clear. May not always be exemplified. Geographical terminology is well used. Good standards of quality of written communication.
Level 2	5–8	Some structure in account that describes some aspects of the role of aid. Some use of geographical terminology. Satisfactory quality of written communication.
Level 1	1–4	Lacks structure. Some basic descriptions about aid. Limited use of geographical terminology. Variable standards of quality of written communication, sometimes with poor grammar, spelling and punctuation.

Student answer

Aid is a simple term for a complex range of types of help. Aid can come in the form of official development assistance or ODA. This is finance for development that can be given bilaterally (country to country) or multilaterally — where countries like the UK contribute to donors such as the World Bank or UN. Aid can also be given via an NGO such as Wateraid or Oxfam. Aid could be money, or goods or even technical personnel. **a**

It is important to recognise that some aid is not given to help narrow the development gap at all. Emergency aid can be bilateral, multilateral or given by an NGO but it is given to stabilise a situation during a disaster and help a country or region return to normality. Following the 2010 Haiti earthquake \$3–4 billion was pledged in emergency aid but most of this will go to rebuild the affected regions, not raise Haiti up the development ladder. **b**

Bilateral aid is often criticised for being 'tied'. This means that the donor country attaches 'strings' to the aid. These usually mean the receiving country has to buy goods and services from the donor with the aid money. A famous example is the way the Pergau HEP dam in Malaysia was funded by UK bilateral aid but the aid was dependent on Malaysia buying British military equipment and the dam was built by Balfour Beatty, a UK company. While Malaysia did benefit from a new source of electricity, the UK seems to have benefited just as much. **c**

About 30% of aid is multilateral and 70% bilateral. Multilateral aid is often given in a top-down way to fund large-scale projects such as dams, ports and air ports. This type of aid might help develop essential infrastructure to help a country develop. Often it is focused on economic development but sometimes this is at the expense of social development. People might be ignored when large-scale development projects are planned. In theory, as a country develops, the benefits of the aid should 'trickle down' to everyone but it is not always clear that this happens. The $900 million World Bank funded Mumbai Urban Transport Project was criticised for failing to re-house 17,000 people moved out of the way for bridges, roads and railways. **d**

It is often said that the most effective way to bridge the development gap is to focus on small-scale, local projects that actually help individual people develop and improve their quality of life. Many NGOs work in this way. This type of aid has the benefit of being low cost and community based, but often it has a small footprint so the total number of people helped is small. Wateraid has helped 20 million people gain access to safe drinking water since 1981, but there are still 900 million people worldwide without it. Wateraid provides small-scale hand-pumps and wells which improve people's health and well-being but don't necessarily improve their economic development. Some small-scale NGO-led aid such as Grameen Village Phone does focus on income generation development. Grameen provides micro-credit loans to women in Bangladesh to buy mobile phones and earn a living from charging people to use them. This extra income could be used for clean water, food education and health care. **e**

Overall, aid can help bridge the development gap. It is often misused and often suffers from corruption. There is a place for aid from a variety of organisations but aid can't be expected to help everyone as the amount of aid is always going to be limited by the generosity of donors. **f**

e **14/15 marks awarded (grade A).** This answer deals with a wide range of types of aid, supported by examples. It does evaluate as it examines the pros and cons of different approaches. **a** This is a very good introduction with an extended definition of aid and its many types; it shows the student has a good understanding of aid. **b** This is a good evaluative example, as it makes clear that not all aid is designed to bridge the development gap. **c** This example shows how aid can sometimes be misused and its aims might not be clear, or wholly about development. **d** This section is evaluative as it identifies the potential economic benefits of multilateral aid, but also the social drawbacks, and provides a useful example as support. **e** This is a good evaluation of the pros and cons of NGO aid; it avoids the 'small is always beautiful' trap and points out that NGOs cannot help everyone and often have a narrow focus. There are good supporting examples. **f** This is a sound conclusion which recognises the limits of aid as well as the fact that it can help.

Question 6 **The technological fix?**

Figure 6 The 'technological breakfast' as consumed by business people in the developed world

(a) Using examples, suggest reasons why this technological breakfast is consumed in some places but not others. (10 marks)

ⓔ Start by thinking about 'some places and not others' as you need examples of contrasting places (different levels of development, political systems) and a range of reasons why some people can access the technology in Figure 6 but others cannot.

Mark scheme and levels

ⓔ In some parts of the world, such as sub-Saharan Africa, almost none of the technology needed for the breakfast would be available due to lack of access to electricity. The breakfast depends on wealth, which determines access to technology to some extent. Global transport technology allows people in the developed world to import some goods, e.g. coffee. Developed countries can afford internet access, but it may also be unavailable (North Korea) or restricted (China) due to political reasons. Wealth explains the use of the Blackberry, although it would be of little use in areas of the world where literacy levels are low. In much of the developing world the food element of the breakfast might be available, but not refrigeration, and fuel wood would be used for cooking rather than electricity. A lack of technology transfer and the cost of royalties could restrict access to the ICT elements of the breakfast in some locations.

Level	Mark	Descriptor
Level 3	8–10	Detailed commentary on several aspects of the figure, which moves well beyond wealth. Answer related to examples. Explanations are always clear. Geographical terminology is used with accuracy. Grammar, punctuation and spelling errors are rare.
Level 2	5–7	Some suggestions that are largely wealth related but other factors are suggested, but not in depth. Focus is on technology. Explanations are clear, but there are areas of less clarity. Lacks full range. Geographical terminology is used with some accuracy. There are some grammar, punctuation and spelling errors.
Level 1	1–4	A few generalised ideas linked to wealth/poverty with limited or no mention of technology. Explanations are oversimplified and lack clarity. Geographical terminology is rarely used with accuracy. There are frequent grammar, punctuation and spelling errors.

Student answer

This breakfast is only available to very few people everyday. It might be a business person in New York, London or Hong Kong. It is an expensive, globalised breakfast. **a**

The person eating this breakfast is technologically connected via their Blackberry and laptop. This means they must have a high income to buy the devices and pay for their connections. This is likely in the developed world but incomes will be too low in the developing world. In rural sub-Saharan Africa both a wi-fi signal and electricity connections are very rare as the money to develop these technologies is not present. Even in some isolated rural parts of the UK broadband and mobile phone signals are not present. **b**

In North Korea, people could have internet connections and mobile phones but the hard-line communist government prevents this to stop people gaining access to information from the rest of the world. **c**

In the developed world, people can afford a range of foods which are imported from all over the world, often using refrigerated container ships. They can also afford labour-saving devices like toasters. In the developing world the breakfast would be cooked on a dung or wood fire. **d**

In some places such as China and India some of the technology might be available but not all of it. For instance in rural India mobile phones are quite common due to technological leapfrogging, but electricity is not. **e**

e 9/10 marks awarded (grade A). This is a very well structured answer, with a brief introduction (no more than is needed for a 10 mark data stimulus question) and summary. The answer works its way around the Figure, providing detailed explanations linked to the availability of technology. Examples of specific places are used and the answer goes beyond the simple idea of rich versus poor. **a** This introduction is good, as it recognises that the breakfast is actually for a select few, and uses the term 'globalised', which shows understanding. **b** This section recognises that some people have the technological connections in the Figure, but others don't. It goes beyond the global North–South idea and makes a valid point about the UK, as well as naming a specific area of the developed world. **c** A good explanation of the situation in North Korea. **d** This is a little over-generalised, but the basic idea is correct. **e** This is a good example, which makes a very valid point that some parts of Figure 6 might not be available to some people.

(b) To what extent do different technological solutions to the same problem produce different impacts? (15 marks)

ⓔ Begin by identifying at least two problems such as global warming and the need for more food. Then decide on two different solutions for each such as GR and GM crops for the food problem. You should then consider the impacts (positive and negative) for each solution and come to a conclusion as to whether or not they are different.

Mark scheme and levels

ⓔ Technologies often fall into particular categories depending on their scale, costs and level of technological sophistication. Some technologies have particular aims, such as sustainability (appropriate technology), meeting basic needs and social equity (intermediate technology) or economic development (mega-engineering). Classes of technology include high-tech, appropriate and intermediate. Technologies can be applied in top-down ways or in bottom-up ways. Examples that might be used include the provision of clean water supplies. These could be achieved using desalination, tube wells, rainwater harvesting, large dams etc. The different technological approaches would have different impacts in terms of energy use, the amount of land used and cost. They might all supply clean water, but in very different quantities. Social impacts could differ as smaller-scale technologies might be more likely to involve local control and decision making, and may be more socially equitable. A large range of other examples might be used, such as farming (GM, Green Revolution or organic approaches) and energy (renewable, nuclear, fossil fuels). Good answers would recognise positive and negative impacts, as well as the range of impacts — social, economic, environmental.

Level	Mark	Descriptor
Level 4	13–15	Structured response focused on impacts; detailed use of real examples which contrast in approach. Provides an overview. Explanations are always clear. Geographical terminology is used with accuracy. Grammar, punctuation and spelling errors are very rare.
Level 3	9–12	Some structure in response which uses some detailed examples of impacts. Less clarity on technology and differences in approach. Explanations are usually clear. Geographical terminology is used with accuracy. Grammar, punctuation and spelling errors are rare.
Level 2	5–8	Descriptive account with some details of various technologies and their impacts, but poorly related to the question at times. Explanations are sometimes clear but there are areas of less clarity. Geographical terminology is used with some accuracy. There are some spelling, punctuation and grammar errors.
Level 1	1–4	A few generalised ideas that describe some technologies and one or two impacts. Explanations are oversimplified and lack clarity. Geographical terminology is rarely used with accuracy. There are frequent grammar, punctuation and spelling errors.

Student answer

One way of solving human problems is to apply a technological fix. This is when technology is developed to solve the problem. Usually the technology will have impacts. These are called externalities and they can be positive and negative. **a**

The key problem facing humans today is global warming. This is caused by there being too much carbon dioxide in the atmosphere. The solution to this problem is simple in some ways because all that needs to happen is for us to reduce carbon dioxide emissions. However, there is a lot of debate over which technological fix to use. **b**

Sulphate aerosols are one solution. This geo-engineering fix would pump sulphur dioxide into the air which would prevent some sunlight reaching the earth's surface and counteract global warming. Ships or smoke stacks could be used to release the sulphur dioxide. People in favour of this argue that it would prevent global warming without humans having to change their lifestyles (by adopting attitudinal fixes). Those against say it could cause huge damage to ecosystems by increasing acid rain.

The impacts of using renewable energy are very different. Wind and solar power would actually reduce carbon dioxide emissions, but they are more expensive than fossil fuels and cannot be used everywhere. This means the impact on some people could be more expensive energy and less reliable energy. **c** These two solutions have very different impacts, especially for the environment. Sulphate aerosols actually increase pollution whereas renewables reduce it. Pollution could affect humans by creating a dirtier environment. Renewables are cleaner but possible more costly. **d**

Lack of water is another problem, with up to 2 billion people lacking a clean, reliable water supply. **e** There are many water technologies. Desalinisation is an energy-intensive way of removing salt from seawater used extensively in the Middle East. It provides humans with fresh water but is expensive and increases carbon footprints. This is a classic high-tech fix. Water harvesting technology such as pumpkin tanks uses cheap, renewable, intermediate technology to collect rainwater in the wet season and conserve it for use during dry spells. These two technologies are at opposite ends of the technology spectrum but they both provide clean water. The impacts of both are the same in terms of water, but different in terms of the resources they use. **f**

Problems can be solved with a wide range of different types of technology. All of these will have social, economic and environmental impacts. The best solutions are those with the fewest negative impacts and most positive impacts. **g**

e 13/15 marks awarded (grade A). This is a good answer. It uses a range of examples, with some details, and recognises positive and negative impacts. It shows a good understanding of technology and addresses the 'to what extent' part of the question. With more detail and a firmer summary/conclusion could gain maximum marks. **a** A good introduction, using geographical terminology and focused on key words from the question, i.e. impacts and solutions. **b** The problem is clearly stated. **c** Two solutions are outlined here in some detail, showing good understanding. **d** This summary addresses the 'to what extent' part of the question by directly contrasting the impacts. **e** This is a useful fact. **f** This is a good contrast between two very different examples, and it includes the assessment that the end result of both is clean water, but also highlights differences in terms of resource consumption. **g** This conclusion is sound; it recognises positive and negative impacts, although it might have linked back to the examples used.

Knowledge check answers

1 Used nuclear fuel can be reprocessed to make new fuel rods, which is an example of recycling.

2 The Middle East supplies over 30% of global crude today; this is likely to exceed 60% in the future.

3 The BRICs all have rapidly growing economies and increasing numbers of consumers, with soaring demand for energy.

4 Choke point shipping lanes could be easily disrupted by terrorism, conflict or piracy threats, which could restrict the supply of crude to other regions.

5 Developed world consumers are heavily reliant on energy for cars, heating, lighting etc. and spend a high proportion of their income on energy, which is seen as a necessity.

6 The main arguments are environmental, i.e. the destruction of ecosystems and possibly water pollution, plus the relatively high cost of some of the resources.

7 OPEC is an economic grouping of major oil exporters. By agreeing how much oil to produce, OPEC can influence the world price of oil, so is a significant player.

8 After peak production, oil supply will drop and as the resource becomes increasingly rare the price of oil will rise. Long before the 'run out' date, oil will become very expensive.

9 Nuclear power has a number of problems, including disposing of high-level waste, high initial costs, being difficult for the public to understand, and being associated with high-profile disasters (e.g. Chernobyl) as well as nuclear weapons.

10 China needs oil to fuel its rapid economic development and Africa has under-developed oil resources, often in countries with little western involvement, such as Sudan.

11 Probably not, as the USA is a superpower but has issues with energy security. Of the BRICs, Russia and Brazil are energy secure, but China and India are not. Economic development, cultural influence, military power and international influence are needed as well.

12 Capital and technology are needed to provide sufficient water of good quality. Rapidly growing populations, especially in megacities, create huge demands. Poorer peoples are less able to cope with climate change.

13 There is a huge demand (China and India contain over one third of the world's people). As these superpowers emerge, this demand for water increases, i.e. for agriculture, industry and domestic use.

14 Many are areas of physical or economic water scarcity, with poor access to improved water supplies, and a low capacity to improve them.

15 Use of groundwater supplies is hard to monitor, and the supplies are poorly mapped, so providing equitable sharing between users is complex. No legal framework has yet been developed by the UN. Many are over-abstracted and diminishing rapidly.

16 National: conflicts between agricultural use for irrigation by commercial farmers and increased demands for city use, e.g. Las Vegas/Phoenix. Native rights issues.
International: in spite of an agreement, USA takes 90% of the water and In dry years (e.g. La Niña) Mexico receives very little water (downstream of the USA).

17 Eutrophication results from farmers over-dosing fields with nitrogen/phosphates, leading to excess fertiliser running into water courses, polluting water supplies and disrupting aquatic ecosystems.

18 Sustainable water use — because more careful management of water resources will allow water use for farming to slowly increase and balance food supply with growing demand.

19 Intermediate technology involves technology that is cheap to install, run and use, such as rainwater harvesting, tank storage or basic wells. High-technology is usually large scale and high cost, such as mega dams, or desalination plants.

20 Integrated schemes treat water resources holistically, with the basin managed by the community. Sustainable water use includes all aspects of policies that conserve the water for future generations (efficiencies) in a way that does not damage the environment.

21 There are many examples, including Mediterranean Spain, southeast England, Andean Republic (glacier melt), much of the Middle East and parts of sub-Saharan Africa and South Africa.

22 Species that are unique in ecological terms, i.e. found nowhere else.

23 It is difficult to establish boundaries and less is known about the value of biodiversity in them. Until recently they were perceived as less under threat. The Australian total is high because of the huge Great Barrier Reef Marine Park.

24 Destruction involves total removals of the forest, e.g. rainforest for timber or growth of soya beans or oil palms for biofuel. Degradation occurs when the forest is damaged, for example by shifting agriculture (it grows back as lower-value secondary forest) or by acid rain pollution.

25 The population (e.g. of fish) will be threatened by over-harvesting, especially if all are extracted including juveniles. The population will therefore decline at one trophic level, so disrupting the rest of the food chain.

26 Advantages — hunting and/or culling the population means that carrying capacity will not be exceeded. Income can be earned from 'trophy hunting'.
Disadvantages — hard to get the level correct. Tempting to over-kill, so could damage the extremely valuable game-viewing tourist industry and also habitat.

27 Hotspot — Myers; eco region — WWF

28 The main ones are WWF, Greenpeace, RSPB, but there are also numerous small-scale specialist NGOs (e.g. Rainforest Trust).

29 Several small reserves can be easy for a community to police and manage locally, and can cover a range of ecosystems. However, to allow corridors of movement (N/S) as climate change impacts occur, increasingly large-scale reserves, often internationally managed, are now considered more successful.

30 Site of specific scientific interest is a UK protection designation for areas of high-value ecological interest (e.g. Oxwich Nature Reserve). For some activities, visitor permits are required.

31 The USA has the world's largest economy for a single nation, the world's greatest military expenditure and the largest TNCs of any nation (China has the largest population and highest carbon emissions).

32 A bi-polar world, with the capitalist USA and communist USSR as the two opposing superpowers. This ended in 1990 with the collapse of the USSR.

33 Wallerstein's theory is a better fit today, as there are arguably three geopolitical groups of countries — the developed powers such as the USA and EU, the BRIC emerging powers and the developing world.

34 India lacks infrastructure such as roads, airports and ports and has problems supplying energy to rapidly growing cities and industry. Poverty is much more widespread in India compared with China.

35 This is indirect control of developing countries' economies and politics via trade terms, trade agreements, aid and international decision making. It contrasts with direct colonial control.

36 Africa, which accounts for only 2% of world trade.

37 Cultural globalisation, especially Americanisation, is rejected because it is seen as eroding local cultures, beliefs and traditions.

38 Internet use grew by over 1000% in China between 2000 and 2008.

39 The BRICS have per capita incomes ranging from $1000 to $9000 and very different ecological footprints and population growth rates; in short they are very different countries.

40 SWFs are government-run investment funds that are used to buy assets like companies or property in foreign countries; as such they can lead to foreign government ownership of assets in another country.

41 In a future multi-polar world there would be many, broadly equal superpowers based in different regions (a more regional world than today), for example China, India, the EU, Brazil and the USA.

42 The Brandt line is a concept developed by Willy Brandt (Mayor of Berlin and leading German politician). The line was drawn to divide the world economically into the rich North and poor South, and this division was subsequently known as the North–South divide. Note this is not simply the two hemispheres.

43 Growing involvement of the developing countries in their future — not the first world developing the third world.
Involvement of local people in grass-roots development, as opposed to top-down colonial models.
Movement towards development as an all-round process, not just economic development.

44 There are eight MDGs — completion date 2015.

45 Combination of generally high birth rates (2.0% per annum) increased by the fact that generally young people migrate to cities, so leading to even higher birth rates.

46 (a) Maturing
(b) Consolidating
(c) Immature

47 → Traditional society → Pre-conditions for take-off → Take-off → Drive to maturity → Age of high mass consumption

48 Cuba. Some argue for Vietnam, or even Venezuela.

49 The distinction can be blurred, but aid includes gifts, grants and low-interest loans of money, goods and people. FDI involves expenditure by a TNC or government wealth fund, with the expectation of financial returns for the donor.

50 Emergency aid, for example after a major famine or disaster, comes from NGOs such as Red Cross/Red Crescent, from particular countries, or even from the UN.

51 (a) WTO
(b) IMF
(c) World Bank

52 Drugs, for example, cost millions to develop, so the pharmaceutical companies apply for patents. These give them the rights to control production and distribution for various designated drugs.

53 Fairtrade is ethical as it involves MEDC companies guaranteeing farmers' groups fair prices for their products (above world market prices) so they can enjoy a better quality of life. The consumers pay higher prices knowing that they have been ethical.

54 UK consumers pay £100–200 per year for broadband, and a little more for high-speed/unlimited access; this is much less than in many developing countries, where incomes are much lower.

55 Subsistence farmers produce food primarily to eat, selling any surplus they manage to produce; most use very basic technology and are not protected from the impact of drought or pest infestations.

56 Drug companies may feel that they are being 'cheated' of legitimate revenue if developing world companies copy their patented drugs.

57 Barriers could include lack of infrastructure (reliable electricity and water supply, transport), lack of skilled workers, lack of component and resource suppliers and even lack of economic and political stability.

58 Small, mobile technologies are often involved in leapfrogging because they do not need to be connected to a pre-existing network or infrastructure; communication and renewable energy technologies are good examples.

59 Using the polluter pays principle (PPP), the person or organisation causing pollution should pay for it — often as a tax, or by paying for the technology to clean material before it is released into the environment.

60 Artificial global dimming uses one form of pollution, i.e. sulphur dioxide, to combat another form of pollution, i.e. carbon dioxide. To most environmentalists this is adding to the pollution problem, not solving it.

61 Wind power produces no carbon emissions, unlike coal, although some are produced during turbine manufacture. Wind is cleaner than coal, but also less reliable.

O

P

R

S

T

U

W